LION HUNTING & OTHER MATHEMATICAL PURSUITS

A collection of mathematics, verse and stories by
RALPH P. BOAS, JR.

LION HUNTING & OTHER MATHEMATICAL PURSUITS

A collection of mathematics, verse and stories by
RALPH P. BOAS, JR.

GERALD L. ALEXANDERSON
Santa Clara University

DALE H. MUGLER
University of Akron

EDITORS

THE MATHEMATICAL ASSOCIATION OF AMERICA
Dolciani Mathematical Expositions Volume 15

THE
DOLCIANI MATHEMATICAL EXPOSITIONS

Published by
THE MATHEMATICAL ASSOCIATION OF AMERICA

Library of Congress Catalog Card Number 94-078313

Complete Set ISBN 0-88385-300-0
Vol. 15 ISBN 0-88385-323-X

Printed in the United States of America

Current printing (last digit):
10 9 8 7 6 5 4 3 2 1

The DOLCIANI MATHEMATICAL EXPOSITIONS series of the Mathematical Association of America was established through a generous gift to the Association from Mary P. Dolciani, Professor of Mathematics at Hunter College of the City University of New York. In making the gift, Professor Dolciani, herself an exceptionally talented and successful expositor of mathematics, had the purpose of furthering the ideal of excellence in mathematical exposition.

The Association, for its part, was delighted to accept the gracious gesture initiating the revolving fund for this series from one who has served the Association with distinction, both as a member of the Committee on Publications and as a member of the Board of Governors. It was with genuine pleasure that the Board chose to name the series in her honor.

The books in the series are selected for their lucid expository style and stimulating mathematical content. Typically, they contain an ample supply of exercises, many with accompanying solutions. They are intended to be sufficiently elementary for the undergraduate and even the mathematically inclined high-school student to understand and enjoy, but also to be interesting and sometimes challenging to the more advanced mathematician.

DOLCIANI MATHEMATICAL EXPOSITIONS

1. *Mathematical Gems*, Ross Honsberger
2. *Mathematical Gems II*, Ross Honsberger
3. *Mathematical Morsels*, Ross Honsberger
4. *Mathematical Plums*, Ross Honsberger (ed.)
5. *Great Moments in Mathematics (Before 1650)*, Howard Eves
6. *Maxima and Minima without Calculus*, Ivan Niven
7. *Great Moments in Mathematics (After 1650)*, Howard Eves
8. *Map Coloring, Polyhedra, and the Four-Color Problem*, David Barnette
9. *Mathematical Gems III*, Ross Honsberger
10. *More Mathematical Morsels*, Ross Honsberger
11. *Old and New Unsolved Problems in Plane Geometry and Number Theory*, Victor Klee and Stan Wagon
12. *Problems for Mathematicians, Young and Old*, Paul R. Halmos
13. *Excursions in Calculus: An Interplay of the Continuous and the Discrete*, Robert M. Young
14. *The Wohascum County Problem Book*, George T. Gilbert, Mark I. Krusemeyer, Loren C. Larson
15. *Lion Hunting and Other Mathematical Pursuits: A Collection of Mathematics, Verse, and Stories by Ralph P. Boas Jr.*, Gerald L. Alexanderson and Dale H. Mugler, Editors

CONTENTS

INTRODUCTION

Occasionally over the course of time, a person with the ability to do high quality research in mathematics also stands out in other ways, contributing to mathematics and the mathematical community in many areas. Ralph Philip Boas, Jr., was such a person: mathematician, author, editor, teacher, and administrator. This collection of his work represents an assortment of many of his lighter mathematical papers along with verse, stories, anecdotes, and recollections. In addition to highlighting his many different abilities, we hope that this volume will stimulate the reader in the way in which RPB intended to stimulate the readers of his articles when they were first published. The mathematics remains fresh and the comments on teaching are as cogent today as when he wrote them. The recollections and anecdotes included here, with a few exceptions, have not appeared in print before. These will provide the reader with a personal glimpse into the life of a mathematician who moved in the circles of the great mathematicians and scientists of his day, such as G. H. Hardy and many others. The stories have tantalizing depth and the verses offer imaginative and insightful views of many aspects of the life of a mathematician.

This volume, then, is by no means a complete set of collected works of Boas, but a collection aimed at the lighter side, both stimulating the reader with accessible mathematics or mathematical methods in teaching as well as offering a view of the development and experiences of an important figure in the mathematical community of the recent past. Some readers may notice that a few of their favorite Boas articles are missing; we have tried to avoid reprinting pieces that appear in other MAA collections.

The organization of this book begins with a glimpse of RPB's personal history and some comments on his life by some who knew him well. Next, there is a section devoted to the legendary piece on the "mathematics of lion hunting" and here we have chosen to include several contributions by other authors who were inspired by the original article. Sections that follow include some devoted to a particular mathematical topic, interspersed with others that contain anecdotes and verse. The order in which verse and anecdotes are presented reflects our aim to draw together either similar topics or events from a particular time or place. Although we have labeled one section with the title "Literature," we have chosen to include short stories in other sections in which they seemed more appropriate, such as the story "The Rose Acacia" in the section on infinite series.

The bibliography at the end of this collection is intended to be a complete listing of RPB's mathematical articles and books, although many reviews (including hundreds in *Mathematical Reviews*) as well as other types of writings are not included in this list. The reminiscences of Boas were either solicited by us or offered to us and we gratefully acknowledge these contributions.

ACKNOWLEDGMENTS

Acknowledgments for those who helped put this collection together must begin with the Boas family, particularly Mary L. Boas and Harold P. Boas, who provided many pieces of information and details that we would otherwise surely have missed. The help continued right up through proofreading.

The person who had the original, creative idea of putting such a book together was Donald J. Albers, Associate Executive Director of the Mathematical Association of America. Without his initiative, it is doubtful that this book would exist, though the material might have appeared in other forms. Prior to his death Ralph P. Boas, Jr. himself communicated with Albers and us on a number of occasions about the initial development of this book.

Rebecca Jackson was a helpful assistant to us in the preparation of the manuscript. And our colleague, Leonard Klosinski, has provided valuable service in the preparation of the index. We are grateful to Deborah Tepper Haimo, Philip J. Davis, Carl L. Prather, and Antoinette M. Trembinska for their touching reminiscences of their contacts with Ralph Boas, and especially to Frank Smithies for his charming essay and for his sharing with us some amusing pictures dating from Boas's visit to Cambridge, one of which has provided the lion that appears throughout the book. We are grateful to the excellent production staff for their fine work in their making the project such a pleasant one for us: from the MAA, Elaine Pedreira and Beverly Ruedi; from Integre Technical Publishing Co., Inc., Donald W. DeLand and Leslie Trimmer; and from Teapot Graphics, John Johnson.

Gerald L. Alexanderson
Santa Clara University

Dale H. Mugler
University of Akron

RALPH P. BOAS, JR.*

Ralph Boas calls himself a "quasi" mathematician because, although he has been a mathematician throughout his professional career, he has done many things besides mathematical research. He has been an editor (Mathematical Reviews and the American Mathematical Monthly), a translator of Russian mathematics, a longtime department chairman at Northwestern, as well as a mathematical versifier and humorist. (See, for example, his work under the pseudonym H. Pétard on mathematical methods of catching lions.) He is married to a theoretical physicist, with whom he has had three children, and has always been dedicated to a two-career marriage. "Real mathematicians," he says, "do only mathematics." He prides himself also on "felling a tree so that it falls where I want it to." Following is his "self-profile of a quasi-mathematician."

I was once having dinner with a group of scientists who were reminiscing about how they had come to take up their professions. One (and only one) of the mathematicians had become a mathematician for a logical reason: he had taken an aptitude test at the age of eleven, had been told that he was best suited to being a mathematician, and so a mathematician he had become. I myself drifted into the field for no better reason than that I was too clumsy to be a chemist and happened to be farther along in mathematics than in anything else.

My parents had both graduated from Brown University (with Phi Beta Kappa keys) and gone on to master's degrees. They were married in 1911, and my father took a job in the English department of Whitman College in Walla Walla, Washing-

More Mathematical People (edited by D. J. Albers et al.), Harcourt Brace Jovanovich, Boston, 1990, pp. 23–41. Reprinted by permission.

ton, where I was born in 1912. We moved around quite a bit in the next few years, and my sister, now Marie Boas Hall, who was to become a well-known historian of science, was born in Springfield, Massachusetts, where my father was teaching English at Central High School.

Sometime after my eighth birthday, my parents decided that it was high time I had some formal schooling. My mother escorted me to a nearby grade school and explained to the principal that I could read. The principal sent her home, saying that as soon as he had time he would see where I should be placed. When I got home, my mother asked, "What grade did they put you in?" I didn't know, but I said we had had to turn in some written work and we had put "6A" on it. My mother was quite alarmed, but I stayed in the sixth grade. School life was rather difficult, because I was not only two years younger than my classmates but also small for my age.

Both of my parents were gifted teachers. My father also had a talent that I wish I had inherited: he could sit through a committee meeting, never open his mouth, and at the end of the meeting say, in effect, "This is what you want to do," and have everybody agree. In his thirties he published some articles in the *Atlantic Monthly*, but by the time I was old enough to understand what he was doing, he refused to write any more articles, although he was always suggesting fascinating ideas for articles. He did write, alone or in collaboration, a number of textbooks. He and my mother collaborated on a biography of Cotton Mather, and my mother wrote several other biographies. Consequently I learned at an early age to read and correct proofs; this is a skill that I have had many occasions to use.

TECHNIQUE BEFORE THEORY

I got a good deal of my education from browsing in my parents' extensive library, in particular from the readings for college students that are the English professor's analogue of the algebra and calculus books that arrive in the mathematician's mail. I cannot recall that either parent made any effort to teach me how to write effectively. Any skill that I now have was learned by practice.

I have only vague memories of junior high school, but I remember learning some Latin and a good deal of formal algebra. In particular, we had worksheets with long lists of equations like $2xy^2 = x^3/y$, to be solved mentally for y, and fast. I developed a proficiency in simple algebra that lasted for a long time and has been very useful. My mother gave me her college algebra book. I learned from it how to solve word problems, although I remember distinctly that I never really understood them. I could do only the problems that followed the pattern of the examples in the book. Somewhat later my father gave me a table of logarithms, and I rapidly learned the techniques of using it; but, again, I didn't understand why logarithms worked. I still find that understanding tends to come after learning formal procedures, although according to conventional wisdom it should be the other way around. My view is that if you have a firm grasp of technique, you can then concentrate on theory without having to stop and think about the technical details. Perhaps that is why I have always liked languages, where practice generally precedes theory.

When I was twelve, my father became a professor of English composition at Mount Holyoke College. There was a kind of informality at Mount Holyoke and later at Wheaton that I believe still persists at small liberal arts colleges. Students were always dropping in on my parents (who lived close to the campus) to discuss their work or their personal problems. The students were, of course, all women and they tended to treat me like a younger brother or nephew until I was in college myself. Perhaps that is why I have never felt intimidated by women.

The principal of my high school was an enthusiast for mathematics and gave a course in solid geometry after school for me and another student. It is interesting that when I graduated in 1928, I knew about as much mathematics (I have been told) as had been required for a mathematics major at Brown twenty years earlier.

In high school I was probably more interested in languages than in mathematics. In my senior year my father asked Mount Holyoke to let me audit a class in first-year Greek. It was understood that I would eventually study the same languages that my parents had studied: French, German, Latin, and Greek. After that, they said, I could go on to more exotic languages.

I did not find high school very demanding, and I expect that it was my parents who encouraged me to do various extracurricular things. For several years I helped Miss Hooker, a retired biologist, raise Buff Orpingtons (a rather specialized variety of chicken). I also worked as an assistant to the librarian of the South Hadley Library and learned how to catalogue and mend books. Miss Gault, who worked in the reference department of the Mount Holyoke Library, took me in hand one year and taught me how to use reference materials.

I graduated from high school two months before my sixteenth birthday. My parents thought (correctly) that I was too immature to go away to college, so my father obtained permission for me to audit classes at Mount Holyoke in Greek, German, and calculus during the next academic year. Our calculus text was a combination of analytic geometry and calculus; this arrangement has come and gone several times since then.

When it was time to decide on a college, I first thought of Reed; but my uncle, George Boas, who taught philosophy at Johns Hopkins, pointed out to me (as I have subsequently done myself in similar situations) the advantages of a large university where there is more choice of subjects. Consequently I decided to go to Harvard. I say "decided to go" deliberately, because at that time Harvard was accepting, without examination, all students who had high enough rank in their high school classes.

My mother had done some substitute teaching, but up to the time when I started college she had never held a full-time job. In 1929 she decided, for reasons that were never explained to me, to get one, and she became an associate professor in the English department at Wheaton College in Norton, Massachusetts. A year later my father joined her as head of the same department.

I had had classes in both chemistry and physics in high school. I had enjoyed chemistry but had not cared much for physics. In the late 1920's there was beginning to be popular writing about the new physics, and people like my parents tried to understand in a general way what was going on. Of course my high school course was of no assistance, and even the Harvard physics course (which was rather boring) gave me no more understanding of either relativity or quantum mechanics than I had picked up from cocktail party conversation. What I know about these fields I learned much later, from my wife.

BROKEN GLASSWARE AND MATHEMATICS

At Harvard, since I had some idea of majoring in chemistry and eventually going to medical school, I talked my way into Qualitative Analysis, but my preparation turned out to be inadequate, and that, combined with my record bill for broken glassware, made me give up that idea. That is how I became a mathematics major.

Since I had had a year of analytic geometry and calculus, I thought I ought to be able to handle the next Harvard course, Mathematics 2. My freshman advisor,

In 1933 as a student at Harvard, where things were done in style: "A style long since vanished."

W. C. Graustein, asked me some questions, including the derivative of the sine function (which I didn't know). Since I did know some material that was not in Mathematics 1, he decided that I could learn what I had missed while the rest of the class was learning what I already knew. It worked out all right.

My teacher in Mathematics 2 was E. V. Huntington, who was an early enthusiast for axiomatized mathematics and wrote many articles on axioms for various systems. I knew nothing of that activity, of course, nor of his work on systems for proportional representation. (His system was subsequently adopted by the United States for apportioning representatives to the states.) He also gave a course on Mathematical Methods of Statistics, which I took to replace the second semester of Qualitative Analysis. This course did not pretend to teach any statistics (which was just as well, considering the state of statistics at the time); it was just mathematics. At midterm I had only a C. I said to myself that that would never do if I was to be a mathematics major, so I worked very hard and ended up with an A. Huntington had a number of unconventional ideas—for example, that weight was more fundamental a concept than mass—and he wrote Newton's second law as $F = (W/g)a$.

In my sophomore year I entered the Harvard tutorial system. Each major had a tutor, either a member of the faculty or an advanced graduate student, who was mainly your advisor but also directed your reading. In effect, it was an extra course in the major field, but you did more or less work according to your temperament. In my last two years, my tutor was D. V. Widder, who began by setting me to collect proofs of the fundamental theorem of algebra. This project got me into reading French and German textbooks and then into the periodical literature. (I ended up with a collection of more than thirty proofs; some years later I managed to devise a new one on my own.)

My mathematics course in my sophomore year was called Advanced Calculus, but actually it was more like what we would now call Methods of Mathematical Physics. One semester was taught (from W. F. Osgood's *Advanced Calculus*) by Osgood himself, who used to tell us German jokes. The pace of the mathematics program was leisurely by modern standards; we spent three years on what we now try to cover in two.

A VANISHED STYLE

Harvard did things in style, a style long since vanished. In my sophomore year the first of the new Harvard "houses" were opened, and I was assigned to Dunster House. The dining room had printed menus and waitresses to take orders from them. Harvard examinations were printed from movable type by the University Press, which also produced for each department a brochure containing not only course descriptions but also a list of all the Ph.D.'s produced by the department with the titles of their theses. The course descriptions could be tantalizing. The one for

Real Variables culminated in "the Riesz-Fischer theorem." We students wondered what that might be. When I took the course from J. L. Walsh, it happened one day that he finished a proof just as the bell rang. "That," he said, as he scooted out of the room, "was the Riesz-Fischer theorem."

In my junior year I took the first course in complex analysis; it was the first course based on rigorous proofs and was supposed to separate the mathematical sheep from the goats. I loved it. There were no textbooks for either real or complex analysis. The instructor lectured; the students took notes. I never thought much about this method of instruction—it was just the way things were—but later it came to seem pointless as long as a textbook could be had. The advent of copying machines has made it even more pointless. However, the escalating cost of printed books may lead to the return of the lecture system. At that point we will be where we were before the invention of printing, when the only way to learn a subject was to listen to someone expound it.

It was about this time that I began to watch the current journals as they came into the library. One day I noticed that Harvard was no longer subscribing to the *Recueil Mathématique* (now *Matematicheskii Sbornik*). Since that journal was mostly in French, I could read it, so I was disturbed enough to ask the then chairman of the department why the subscription had been allowed to lapse. (Perhaps fortunately, I have forgotten who the chairman was.) He replied, "Oh, are the Russians doing anything interesting?"

A POSSIBLE INDIC PHILOLOGIST

In my junior year I was taking three mathematics courses and wanted something as a distraction. More or less by accident I settled on Sanskrit. The Sanskritist, Walter E. Clark, provided us with a grammar, a dictionary and a text. After a week spent learning the alphabet (forty-eight letters, which combine with one another in elaborate ways), we started reading, learning the grammar as we went. Some people dislike learning a language this way, but I found it congenial. I kept up Sanskrit for two more years and with encouragement might have ended up as an Indic philologist.

During the summer of 1932 I tried to reproduce the proof of a theorem on Taylor series that I had seen in an old paper by Pringsheim. I couldn't do it, and when I went to Cambridge and looked it up, I found out why: the proof was wrong. This eventually led to my first research paper, the story of which I have written up for the *Mathematical Intelligencer*.

There was a senior comprehensive examination in mathematics at Harvard, about which I remember only that we were asked to prove that the distance to the horizon in miles is approximately the square root of 3/2 of one's elevation in feet. I struggled with this for a long time but could only prove it with 8/5 instead of 3/2. It was only when I was walking away from the examination that it struck me that 8/5 *is* approximately 3/2.

When I graduated in 1933, I was high enough in my class to be awarded a Sheldon Fellowship, which was not for study but for travel. During 1933–34 I wandered around Europe, visiting places I had read about. In the course of my travels I found a proof of Pringsheim's theorem.

In the fall of 1934 I returned to Harvard and began to get into serious mathematics. I had a semester of modern algebra with Saunders Mac Lane and found it thrilling. I thought of specializing in it, but the second semester (with a different instructor) was so dull that I gave up that idea.

I took a course in potential theory with Oliver D. Kellogg when I was an undergraduate, and would have gone into that field except that Kellogg died. There was, however, a legacy from the course. In Kellogg's book there is a rather complicated derivation of the normalization constant for Legendre polynomials. I found a short proof by using recursion relations, which impressed Kellogg. This is now the standard proof but probably only because the time was ripe for it to be found. However, just at that time Pauling and Wilson were writing their book on quantum mechanics at MIT, and I like to think that Kellogg may have told them about my proof; it is given in an appendix to the book.

On a Sheldon Fellowship (awarded specifically for travel), Boas and a group of friends stumbled upon a skeleton during the course of a ramble in Greece.

Boas's thesis advisor, D. V. Widder, lecturing at a symposium in his honor.

Eventually I wrote my thesis with Widder, whose kind of mathematics appealed to me. Before I wrote my thesis, I had written, at Widder's suggestion, a paper on the Hausdorff moment problem. After I had presented my results at a Harvard colloquium, Marshall Stone asked me what I was planning to do next. I told him, adding diffidently that I didn't suppose it was really very interesting. He snapped, "Then why do it?" I took this comment very much to heart. As far as I can remember, I have never since written, "It may be interesting that...." I say firmly, "It is interesting that...," on the grounds that if it interests both me and the referee, then it *is* interesting. In fact, I rather pride myself on never having worked on anything that wasn't interesting, at least at the time.

UNIFORMITY BY THE TAIL

My thesis was something that Widder wanted someone to work out; it was part of the evidence that led Widder and Hirschman to their theory of convolution transforms. Widder (whose ninetieth birthday was celebrated recently) is a rather formal person; it was only many years after I had left Harvard that he came to address me by my first name. It is always startling when he lapses into a colloquialism. During my senior year I once went to him with something I had worked out, and he said, "This shows that you have uniformity by the tail."

I also had to write a minor thesis. This, as far as I know, is a peculiarly Harvard institution. You are given a topic outside your own field and have to produce an essay on it in three weeks (if in the summer) and four (in term time). I drew dimension theory, about which there was at the time only Menger's book besides the periodical literature. There were two rules: you mustn't do research on your topic, and you mustn't repeat any mistakes that were in the literature. (Menger's book was notoriously full of mistakes.) People, even some now eminent people, have failed the minor thesis by getting carried away and doing research on the assigned topic. I think I have never worked so intensely, before or since, but I was left with the comforting feeling that I could learn anything in mathematics with three weeks' warning. In fact, in my first teaching job I suddenly had to prepare a course in probability, which I had never studied before. Years later I looked at my lecture notes and decided that I had done a pretty good job.

For 1937–38 I got a National Research Fellowship and went to Princeton to work with Salomon Bochner. It was a stimulating place because some of the basic theorems in functional analysis (a term that had not yet been invented) were being worked on there.

Bochner had a number of standard responses to any problem you asked him about. They ranged from "I think this is not very interesting" to "I think this cannot be." Once I got "I think this is difficult" and then solved the problem. When I took the result to Bochner, he said, "I think this is trivial." As Widder used to say, "Everything is trivial when you know the proof."

CATCHING LIONS WITH MATHEMATICS

In Princeton I usually had dinner with a group of (mostly) mathematicians. One of the things we talked about was mathematical methods for catching lions. There were many jokes in this vein circulating at Princeton at that time. Below are a few examples:

The method of inversive geometry. We place a spherical cage in the desert, enter it, and lock it. We perform an inversion with respect to the cage. The lion is then in the interior of the cage, and we are outside.

The Peano method. Construct by standard methods a continuous curve passing through every point of the desert. It has been remarked [by Hilbert] that it is possible to traverse such a curve in an arbitrarily short time. Armed with a spear, we traverse the curve in a time shorter than that in which a lion can move his own length.

A topological method. We observe that a lion has the connectivity of the torus. We transport the desert into four-space. It is then possible to carry out such a deformation that the lion can be returned to three-space in a knotted condition. He is then helpless.

Frank Smithies (who was visiting from Cambridge, England) and I undertook to write an article about this interesting field, inventing a few extra methods as we went along. We picked Pondicherry (one of the French enclaves in India) as a pseudonym, spelling it Pondiczery to make it look Slavic; we thought of Pondiczery as being Poldavian like Bourbaki. We submitted our article to the *American Mathematical Monthly* with a cover letter saying that the author, afraid of repercussions, wanted to use the pen name H. Pétard. In an endeavor to establish a reputation for Pondiczery, we imitated Bourbaki by publishing short notes under his name. Later, when I was teaching at the Pre-Flight School during World War II and wasn't supposed to publish anything, Pondiczery wrote a substantial number of reviews for *Mathematical Reviews*.

At Smithies's suggestion I spent the second year of my fellowship in Cambridge, England, to which he was returning. The fellowship was not supposed to allow foreign travel, but I persuaded the authorities to let me go if I paid my own way, which I could do because I had saved enough from my $1600 stipend. In Cambridge I learned quite a lot about England and quite a lot of mathematics. I attended lectures by Hardy, Littlewood, and Besicovitch and also Hardy and Littlewood's conversation class (American: seminar), which met in Littlewood's rooms but always without Littlewood.

One day I went into the mathematical library, glanced at the shelf of new journals, but saw nothing of interest. Smithies came in and asked, "Anything interesting today?" "No," I replied in a disgusted tone of voice, "only the *Proceedings of the Lund Physiographical Society*." Frank went over and picked it up. It turned out to contain Thorin's famous paper on the Riesz convexity theorem and caused a sensation in Cambridge. I now distrust people who want to disregard minor journals.

HARDY'S THREE QUESTIONS

At that time Hardy was an editor of the *Journal of the London Mathematical Society*. He used to tell referees to ask three questions: Is it new? Is it true? Is it interesting? The third was the most important. I would now add: Is it decently written? I think Hardy took that as a given. If he got a paper that was interesting but badly written, he would ask a graduate student to rewrite it. I know, because I did a couple of such rewrites for Hardy, for authors who subsequently became very well known.

I do not remember having thought very much in England about what I was going to do next. Before I had started to worry, I was offered an instructorship at Duke University. J. J. Gergen, who was chairman at Duke, had had me in his course in Fourier series while he was a Benjamin Peirce Instructor at Harvard. He must have been impressed, because somebody told me later that Gergen had worried that Leonard Carlitz and I might form a "Jewish clique"; however, I can't complain about the way Gergen treated me. Actually, Carlitz and I had very different interests and hardly ever spoke to each other.

EVENING OFFICE HOURS

In my first two years at Duke, I slept in a dormitory room but really lived in my office in the physics building. I kept evening office hours, which was popular with the students because, of course, evenings were when they tried to do their homework. Once a student came to see me because he couldn't understand why one problem on his final exam had been marked wrong. It turned out that it had been graded incorrectly; in fact, everybody's paper was incorrectly graded on that problem. We had to re-grade some two hundred bluebooks. The interesting thing was that, at least in my class, only one final grade had to be changed; it was the grade of the student who had complained.

While I was in Cambridge, *Mathematical Reviews* had been founded; when I got home, I found an invitation to be a reviewer. I would have been insulted if I had not been asked, but I couldn't help thinking that reviewing might be a lot of work. In fact, some members of the mathematical Establishment were worried at the time that reviewing might take people away from important research. This has hardly been the case for me. I not only found MR very helpful in my research, but some of my ideas were suggested by papers I reviewed.

REVIEWING AND REAL MATHEMATICS

Since I had few demands on my time at Duke, I reviewed papers rather fast; consequently I got a lot to review. I eventually decided that I needed to be able to read Russian. My method of learning was to notify MR that I was willing to review Russian papers. I promptly received a four-page Russian paper. All that I had was a dictionary and the knowledge (acquired from a Russian-reading acquaintance) that *-ogo* is a genitive ending and is pronounced "ovo." It took me a week to puzzle out that paper. Ultimately I reviewed quite a number of Russian papers, many of which were useful in my fields of research.

My interest in reviewing was probably the first sign that I was not going to be a real mathematician. Real mathematicians, except for a small number of geniuses, don't do anything *except* mathematics. (My classmate Angus Taylor claims that this is what his advisor told him was required for a successful career in mathematics.) I was already getting tired of doing almost nothing except mathematics, and over the years I have drifted into peripheral activities. Although I am fond of classical music, I never learned to play an instrument, and I am hopelessly unathletic. However, I grew up in the country and summered on Cape Cod, so I console myself by being

BOAS' (ALMOST) SECRET CAMBRIDGE ALBUM

At Cambridge young Boas, armed with an unobtrusive and quiet secondhand camera, managed to catch a number of mathematicians unaware before he was "apprehended" by Dame Mary Cartwright.

R. P. B. in 1938 in Cambridge, England.

J. E. Littlewood.

André Weil.

G. H. Hardy.

W. W. Rogosinski.

Dame Mary Cartwright.

A. S. Besicovitch.

able to do some things that my more cultivated colleagues probably can't. I do, for example, know how to sail a boat, shingle a roof, cut grass with a scythe, and fell a tree so that it will fall where I want it to.

MISS LAYNE—GLOWING RECOMMENDATIONS

One day in the spring of 1940, when I was in Gergen's office, he mentioned that he had hired a woman to teach on the East Campus (the women's campus, although some of the women had classes with the men on the West Campus). For no particular reason that I could see, he gave me her (very glowing) letters of recommendation to look at. It was time to reserve seats for the next year's campus concert series, and it occurred to me that if Miss Layne liked symphonic music, it would be a friendly gesture to get tickets for her.

There was a complication, because Miss Layne was going to be a half-time student as well as a half-time instructor. I had grown up around college campuses, and I knew perfectly well that faculty don't date students. However, I chose to identify Miss Layne as a colleague rather than a student. I took her to the concerts, and we went on from there. By the next spring we were engaged, and everybody was surprised. We told our news to Gergen with considerable trepidation, but he raised no difficulties about Mary's keeping her job after we were married.

I had heard, in my parents' home, about engaged couples neglecting their work, but nothing like that happened to us. I did some of my best mathematics that spring; and in the multi-section course, of which we each had a section, where the examinations were graded collectively with results tallied by sections, we were very pleased to find that our two classes came out at the top. Mary actually had more teaching experience than I, and I learned a lot from her.

AHLFORS SHOCKED

We were married in June 1941. The United States entered World War II in December 1941. In 1942 the Navy Pre-Flight School was established in Chapel Hill, five miles away, and several of us local mathematicians applied for jobs there in the hope of not being drafted. We taught three classes a day, made up tests and graded papers. The classes included elementary physics as well as mathematics and went very fast. We were allowed fifteen minutes to teach interpolation, which was rather important in those pre-calculator days. The course included enough spherical trigonometry so that we could explain celestial navigation. I had never studied any kind of trigonometry beyond right triangles until I had to teach trigonometry at Duke. Incidentally, when Ahlfors arrived at Harvard in the 1930's, he professed to be shocked that no spherical trigonometry was taught there.

The day at the Pre-Flight School was divided into three segments: Academic, Military and Physical, which rotated through the day on a two-week cycle. On those days when the cadets had been on a twenty-mile hike in the Carolina sun, followed by lessons on how to kill each other with their bare hands, they were not very alert in mathematics class. Some of them were not very sophisticated in mathematics. One cadet, who had a private airplane pilot's license, was failing mathematics. When he was asked how much gas he would need to carry if he were going to fly two hundred miles at so many miles per gallon, he didn't know whether to multiply or divide. How, the officers asked, was he able to get the right answer? He replied that he did it both ways and took the reasonable answer. They felt that anybody who knew what was a reasonable answer had promise, so they gave him a second chance.

In the spring of 1942 the Navy decided that it wanted no more civilians in the program. Los Alamos was starting, and I interviewed for a job there. Although Hans Bethe, who interviewed me, was very closemouthed, I knew enough to see that they were intending to make what we then called an atom bomb. I didn't get the job, and I have since been very thankful that I didn't. At the same time Harvard was starting a couple of military training programs and recruited me.

It had been understood between Mary and me from the beginning that we were to have two careers, but it hasn't been altogether easy. In the first place, although my parents had both been teaching at the same college for years, there were more anti-nepotism rules in the 1940's than there are now. (It took twenty years to get Northwestern's abrogated, even though it was not written down anywhere.) Second, there was a general feeling (except at women's colleges) against employing women at all. Of course, under stringent circumstances like a war, the prejudice tended to be forgotten—temporarily. Almost as soon as we arrived in Cambridge, the telephone rang. The Widders were putting us up temporarily, and Tufts wanted to know if Widder knew any candidates for instructorships in mathematics. Widder replied that he had one right there, "if you can use a woman." They could, and Mary taught at Tufts for several years. A few years later, when they needed another mathematics instructor, someone said, "Of course, we don't want a woman." The point is that they no longer thought of Mary as "a woman" but as "Mrs. Boas," a respected member of the department. That didn't stop them from firing her when our first child was born, even though by that time she had her Ph.D. in physics from MIT.

At Harvard I taught primarily in the Navy program (as Mary did at Tufts), but I also taught some regular classes. One of these was a Radcliffe class (Harvard was not yet coeducational), which was better than any other calculus class I have ever seen. The Navy classes were very well disciplined, because if the students stepped out of line, they got sent back to the fleet. I had the satisfaction of seeing some sailors who had started with practically no background learn fast enough so that they came out in the end as commissioned officers.

GRASPING

By early 1945 it was clear that the war was coming to an end and that the regular faculty would be returning to Harvard. Fortunately for me, just at that time Willy Feller had decided to leave the part-time editorship of *Mathematical Reviews*. The work had increased to the point that a full-time editor was needed. The American Mathematical Society offered me $4,000 a year. I had been getting $4,800 at Harvard, and I held out for that. The AMS thought that I was very "grasping," but since they had no other candidate, they gave in.

By that time Mary was working for her Ph.D. at MIT. We had a nice apartment in Cambridge, and apartments were scarce, so I commuted: fifty minutes on the train to Providence plus a walk up the hill; I did this for five years. The editorship made me temporarily well known in mathematical circles.

Part of my job at MR was to translate the titles of papers that were not in one of the four canonical languages (English, French, German, and Italian). Of course I made some mistakes, but usually the reviewers corrected me. The worst problem, which occurred in 1947 when the Soviets started publishing exclusively in Russian or regional Soviet languages, was translating Russian titles, since even the papers in the other languages usually had Russian summaries. At one time there was an organization that issued a list of translated Russian titles, but they weren't always accurate. In some cases I could figure out the correct translation only by mentally putting the alleged translation back into Russian and then translating it.

In one case when I translated the Russian summary of a paper in Georgian, I found that it merely said, "A Russian translation of this paper will appear elsewhere." That, and similar experiences, induced me to learn a little (but only a little) Georgian, which is difficult because it belongs to a language family completely unrelated to the Indo-European languages. I have always been charmed by the fact that the Georgian word for "father" is "mama." This controverts psychologists who have convincing reasons why a child calls its mother "mama." (In case you are wondering, the Georgian word for "mother" is "dedi.") The Georgian alphabet seems to have many unexploited possibilities for mathematicians who need new symbols.

MR once received a paper in Gaelic. I looked in the file where reviewers had listed the languages they could read. It included only one person, R. A. Rankin (whom I had met in Cambridge), who claimed to read Gaelic, so I sent the paper to him. The review came back with the note, "I suppose you know who wrote this paper." The author, Rob Alasdair MacFhraing, was Rankin himself in Gaelic disguise.

When I had to cope with a Chinese journal that was wholly in Chinese, I thought that I ought at least to find out who the authors were. I took it to C. C. Lin, and he transliterated the names for me. I asked him what would happen if a name contained a character that he didn't know. He gave me a funny look and said, "I'm supposed to know them all."

RUSSIANS PUBLISH JUNK, TOO

When George Mackey told me that he had been learning Russian and was willing to review Russian papers, I sent him the first one that came along in his field. A few days later he said to me, "Did you know that the Russians publish junk, too?" Up to that time, of course, he had seen in translation only the most interesting Russian papers.

Willy Prager, a leading expert on, among other things, the theory of shells, also read Russian. One of his reviews merely said in effect, "This is a paper on shell theory." I complained because a review ought to say what was in a paper. Prager then told me that he himself knew that much deeper results were known in the United States, but they were classified. He hadn't wanted to give the impression that the Russian results were new, but he couldn't say why they weren't. What occurred to me was that if the Russians were publishing material that we classified, what must their classified material be like?

"WHAT OTHER KIND OF DERIVATIVE IS THERE?"

In the early days at MR we had time to edit reviews. I remember when Feller and I went over Abraham Wald's review of von Neumann and Morgenstern's book, *The Theory of Games and Economic Behavior*. The review had been long delayed.

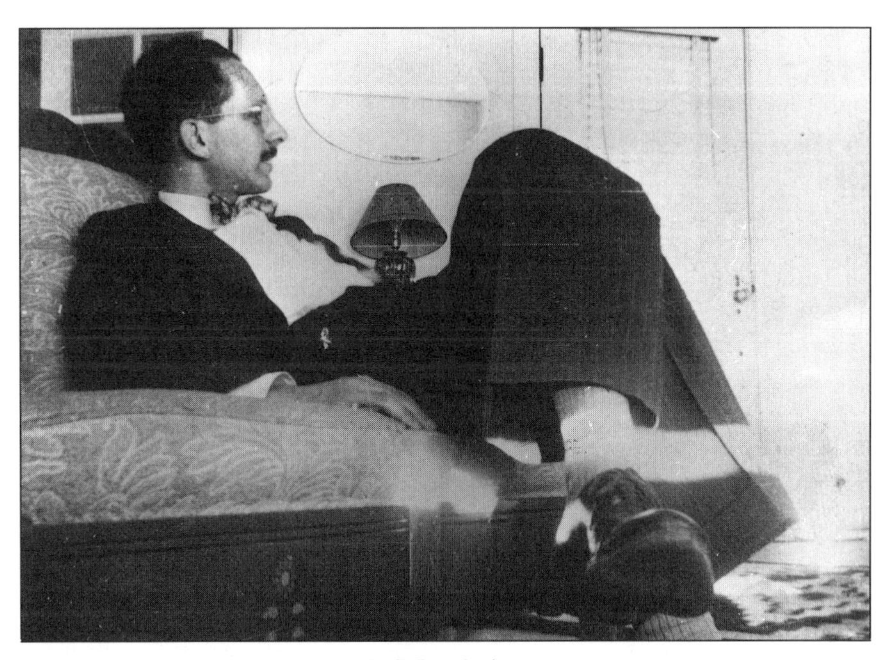

R. P. B. relaxing.

Feller had asked Wald why he had published a paper on game theory before sending in his review. He replied that he had to read only one chapter before writing his paper, whereas he had to read the whole book before he wrote the review. The review was twelve pages long; Feller and I managed to compress it to six pages. For example, Feller deleted "partial" from "partial derivative," saying "What other kind of derivative is there?"

I became editor in 1945, when MR published 280 pages a year, or about two thousand reviews. The staff consisted of one secretary (Janet Sachs) and me. Between the two of us, we did everything outside of actually setting the type and mailing the issues. On Mondays, Janet gave me the journals that had come in during the past week. I assigned the papers to reviewers (there were about three hundred of them, and I very soon knew them, their fields, their idiosyncrasies, and even their typewriters), and the papers went downstairs to be microfilmed. On Fridays they went into the mail.

I had been told that if there wasn't much to do, I needn't come in every day, but that never seemed to happen. One Monday there were only three papers to distribute, but then the mailman came in with a big load of Romanian journals from the war years. After that the work got heavier very rapidly. We had kept up with the German journals quite well, because Otto Neugebauer had been able to arrange for a friend in Switzerland to get them for us, but we hadn't received any Japanese or Italian material. There were few surprises from the war years. The biggest that I recall was Cesari's work on surface area.

We needed more reviewers for Russian papers after the Soviets went all-Russian, and for a while the ones we had were rather overworked. In the late 1940's the American Mathematical Society started translating Russian papers, and for a time that project was run from the MR office.

It was fascinating to have the world's mathematics flowing across my desk, but it got more and more tiring. In 1949 I applied for a Guggenheim Fellowship. At that time (although I didn't know it) Northwestern University was in the process of building a research-oriented department. They wanted a chairman, and I was asked to go for an interview. I remember that I asked the Dean what restrictions there were on the kinds of people I might hire. He replied, approximately, "None now, but two years ago I wouldn't have been allowed to hire you."

In the end I decided that I didn't really want to be a chairman, but Northwestern asked me to join the faculty anyway. Thus I jumped from having been nothing higher than an instructor to being a full professor. The administration worried that since I had been out of teaching for five years, I might no longer be able to teach, so they started me off with a five-year contract. (It was made permanent after the first year.)

With the move Mary needed a job too. It is hard, even now, for a couple to find jobs in the same area, even if not in the same institution. She was eventually hired at DePaul University and rose to a full professorship there. We would have moved to a more congenial location later, but we never found a place where there were two jobs and where we wanted to go.

At that time (1950) the normal course load of a professor was twelve hours a week, although I was given only nine to start with. There was no regular sabbatical system at Northwestern and still isn't, but the University has always supplemented Guggenheim and Fulbright awards and has been generous with leaves of absence. In my case they asked the Guggenheim Foundation to let me postpone my fellowship for a year so that I could start in my new position.

It turned out that it had been a good thing that I had not accepted the chairmanship. I rapidly made myself unpopular with the higher administration by asking them to be more idealistic, in small ways, than they wanted to be. A few years later, after H. T. Davis had retired as chairman, the Dean polled the department about whom they wanted as chairman. Some wanted Walter Scott, who had been acting chairman for a year; some wanted Jean Dieudonné; some wanted me. Although I was still considered unacceptable as chairman, the Dean found a solution. He discovered that the statutes of the University provided that a department could be governed by an administrative committee, which could choose its own front man (to have lunch with the Dean and do everything except make decisions). He called the three of us into his office, gave us our mandate, and walked out. We looked at each other for

Dad Boas reading to Anne and Ralph L.

The Boas family in 1962: Mary, Anne, Harold, Ralph L., and Ralph P.

a while. I knew that Dieudonné wouldn't do anything administrative, so finally I suggested that since Walter had been doing all the work during the past year, maybe it would be a good idea if I took a turn. That was that.

The arrangement had several advantages. In the first place, it allowed me to fit my own teaching schedule around Mary's. It was also ideal in another way: whatever I wanted to do, I had only to convince one of the other two; they were so different from each other that I was sure to get what I wanted from one of them. Things went so smoothly that I was asked to continue for another year. After that year the administration decided that I wasn't so bad after all, and I was appointed chairman for five years. I served two more five-year terms, after which I quit.

One of a chairman's duties is to keep peace in the department. This is not always easy. The chairman can do a certain amount by giving people what they want, but even this has its dangers. H. C. Wang once told me, "The trouble with you is that when I ask you to do something, you do it before I have time to change my mind." Sometimes people have conflicting wants. One day two professors both asked for private conferences, in which each complained bitterly about the other. On another occasion I received a deputation of students who complained about their instructor. They were followed by another group of students who admitted that they were doing something unusual: they wanted me to know what a wonderful teacher they had. Same class, same teacher.

KEEP AWAY FROM THE OFFICE

I found that as chairman it was best to keep away from my office as much as possible. If I was there too much, I found myself looking for things to make me appear busy.

Mary and I eventually had three children, and we made it a principle that there should always be a parent at home if any of the children were there. As I recall, only once in seventeen years was there a complete jam when I had to engage a sitter. In the early days Mary and I shared a study behind French doors. It was an inviolable rule that if either parent was in the study with the doors closed, a child was not under any circumstances to enter the study but was to find the other parent. Mostly, however, we did our work in the study at night.

A POSITION SECRETLY COVETED

As time went on, I came to do less research and more of other things. My involvement with the Mathematical Association of America began when I was invited to a conference of CUPM (the Committee on the Undergraduate Program in Mathematics). I do not have my father's ability to keep my mouth shut. What I said, however, must have seemed relevant, because I was invited to more conferences. I became a member of CUPM, then its chairman, and finally the president of the

With his son, Harold, also a mathematician.

MAA. After that I became editor of the *American Mathematical Monthly* (a position I had secretly coveted).

As editor of the *Monthly* I inherited a year's backlog of articles. After I had worked these off, I tried to keep most of the main articles at a level appropriate for the majority of the readers, who are typically college teachers, but I did accept occasional articles at a higher level on the grounds that the more sophisticated readers of the MAA deserved to have something now and then.

I was able to maintain some control over the style of the articles by editing them, sometimes rather extensively. The ones that needed the most editing, I regret to say, were articles on teaching mathematics, written by educators. The amount of poor grammar and bad spelling in many articles (in whatever field) was quite distressing. One would hope that one's colleagues would write clearer and more accurate prose than reporters for a daily newspaper, but they don't always.

It has been only recently that editors of the *Monthly* have tried to modify it. Although some MAA members resist any change, the editor can change the format to some extent; modifying the content is more difficult, because the editor can do little more than select from what comes in. In my experience inviting people to write articles does not produce very much usable material. One modification that I made has persisted: changing the color of the cover from year to year.

At various times I have served on a fellowship panel for the National Science Foundation and on the Committee on the Advanced Mathematics Examination for the Educational Testing Service. Activities like these give me a feeling, perhaps factitious, that I am doing something socially useful. That is one reason why I am not a real mathematician. The best mathematicians either don't get on boards and panels or don't do any work if they are appointed. Real mathematicians (with a few exceptions) do nothing but research and only enough teaching to justify their professorships. I am more likely to think of myself as primarily a teacher. I was, in fact, reasonably successful as a classroom teacher and as a director of Ph.D. candidates. I spent most of my working time for two years editing two volumes of George Pólya's collected papers. Some of my activities have not turned out too well; in particular, CUPM did not live up to its promise, perhaps because we did not foresee the changes that the computer would bring about and because we expected too much from the teachers.

After I retired, I expected to supplement my retirement stipend by doing things, like translating, that would keep me from being bored. The first year after my retirement, I was asked to teach the first quarter of advanced calculus. At the end of the quarter the department had no more money, but nobody wanted to take the course over, so I taught it without pay for two quarters. There was some static from the administration, but I contended that as Professor Emeritus I was officially entitled to "the privileges of the University," one of which is obviously the right to teach. In the end Northwestern hired me from time to time to do part-time teaching until I no longer had the physical stamina to meet classes.

Boas has a finely tuned sense of humor.

"IF IT ISN'T FUN, WHY DO IT?"

I believe very deeply in the importance of mathematics in general. I am not happy about the dichotomy between "pure" and "applied" mathematics, nor about that between teaching and research. I cannot agree with a past president of my own university, who claimed that only a successful research mathematician can be a successful teacher of mathematics. For many years I tried to be both as well as to carry on with other activities, some of which stemmed from my commitment to a two-career family. I believe that I did an approximately equal share in bringing up our three children and in the essential housework. Naturally enough, as I got older, I no longer kept up with mathematics as carefully as I used to, but I have never completely stopped thinking about mathematical problems, although my ideas tend to turn out to have been anticipated. I have put a good deal of effort into didactic and expository writing. I was nearly seventy when I rescued *Selecta Mathematica Sovietica* from early collapse. I was over seventy when I was invited to become one of the principal editors of the *Journal of Mathematical Analysis and Applications.* At seventy-five I undertook the lexicographical part of the American Mathematical Society's revision of the *Russian-English Dictionary of the Mathematical Sciences,* mainly because nobody else seemed willing to do it.

Some years ago, after I had given a talk, somebody said, "You seem to make mathematics sound like so much fun." I was inspired to reply, "If it isn't fun, why do it?" I am proud of the sentiment, even if it is overstated.

REMINISCENCES OF RALPH BOAS

Frank Smithies
St. John's College
Cambridge, England

Ralph and I first met when I was visiting Harvard in May 1937 to give a talk on singular integral equations. In conversation at tea-time, we discovered a common interest in the little oddities of life, and got onto the subject of long place names; I produced the village in Wales called

Llanfairpwllgwyngyllgogerychwyrndrobwllllandysiliogogogoch

(Church of St. Mary in a hollow of white hazel, near to a rapid whirlpool, to St. Tysilio's church, and to a red cave), and he produced

Chargogogogmanchargogogchaubunnagungamaug,

the name of a lake in Massachusetts, reputed to preserve the terms of a treaty between two local tribes (I fish on my side and you fish on your side and nobody fishes in the middle). Many years later Ralph produced a beautiful New Zealand place name, to wit

Taumatawhakatangihangakoauauotamateapokaiwhenuakitanatahu.

He didn't then provide a translation, but I have since discovered one: "The brow of the hill where Tamatea who sailed all round the land played his nose flute to his lady love."

We also discovered that we were both planning to spend the academic year 1937–38 in Princeton, Ralph to work with Bochner in the University, and myself to continue working with von Neumann at the Institute for Advanced Study; we were both recent Ph.D.'s, holding post-doctoral fellowships. The Institute did not yet have buildings of its own; the mathematicians of both bodies shared (the old) Fine Hall. Ralph was living in the Graduate College and I was having meals there, though I lodged elsewhere.

When Ralph arrived, he soon joined in the usual occupations of our crowd: talking about mathematics, current affairs, or anything else; walking in the countryside, going to the movies, listening to music (and to the radio news at times of crisis), playing ping-pong (our game could hardly be dignified as "table tennis"), taking part in play-readings. Some of the other members of the crowd were Arthur Brown, Ralph Traber, Lyman Spitzer, Hugh Dowker, John Olmsted, Henry Wallman, a physicist named Mort Kanner and a modern linguist named Dick Jameson; other new arrivals that year included George Barnard (like myself, from St. John's College, Cambridge) and John Tukey (from Brown University).

At some time that winter we were told about the mathematical methods for lion-hunting that had been devised in Göttingen, and several of us came up with new ones; who invented which method is now lost to memory. Ralph and I decided to write up all the methods known to us, with a view to publication, conforming as closely as we could to the usual style of a mathematical paper. We chose H. Pétard as a pseudonym ("the engineer, hoist with his own petard"; *Hamlet*, Act III, Scene IV), and sent the paper to the *American Mathematical Monthly*, over the signature of E. S. Pondiczery (we liked the name, and confused matters by spelling it as if it were Polish), who explained that he would prefer to publish the paper pseudonymously. Pondiczery's existence was established by the publication of a short note in the May 1938 number of the *Monthly*, and his name has since appeared in various periodicals; the most recent occurrence that I know of was in the *Mathematical Intelligencer*, 4, p. 2 (1982). The lion-hunting paper was duly accepted for publication, with one editorial alteration: our footnote to a footnote was ruthlessly removed. The paper succeeded beyond our wildest dreams; we gathered afterwards, for instance, that Steinhaus read a translation of it to the mathematical seminar in Lwów; and it has been reprinted in other places.

We also engaged in more serious joint ventures. Aurel Wintner was giving a course of lectures on infinite convolutions and asymptotic distributions; the audience consisted of Boas, Tukey, myself, and C. C. MacDuffee, who was spending the year in Princeton. Boas, Tukey, and I were commissioned to write up the notes of the course; after a little while, Tukey found himself too busy with other things to take an active part, so it fell to Ralph and me to do most of the work, though we consulted John occasionally. The writing was mainly done in Ralph's room in the Graduate College; we refreshed ourselves from time to time with chocolate and a Greek wine called Mavrodaphne, a not too incompatible combination. At the end of the lecture course, MacDuffee, feeling that he too ought to be involved somehow, took the three of us for a day trip in his car up to Lake Hopatcong. And so it was that, when the notes eventually appeared, they were described as being "by R. P. Boas,

Jr., F. Smithies and J. W. Tukey, with the sympathetic encouragement of Cyrus C. MacDuffee."

Ralph and I wrote a joint paper on the relation between the analytic character of a distribution function and the order of magnitude of its Fourier transform; we had independently thought of different ways of tackling the problem, and we incorporated both approaches in our paper, which appeared in the *American Journal of Mathematics*. We also collaborated in giving a course of lectures about the recent work of Kantorovich on partially ordered vector spaces.

Two more Princeton memories. The spelling "ghoti" for "fish" is well known ("gh" as in "laugh," "o"' as in "women" and "ti" as in "nation"). I produced (not original with me) "ghoughphtheightteeau" for "potato" ("gh" as in "hiccough," "ough" as in "though," "phth" as in "phthisis," "eigh" as in "neigh,""tte" as in "gazette," "eau" as in "beau"). Ralph retorted with "ghghgh" for "puff" ("gh" as in "hiccough," as in "Edinburgh" and as in "laugh"). Tucker had published a paper under the title "Degenerate cycles bound"; Ralph concocted a limerick:

A. W. Tucker has found
Results both new and profound,
But none that attract
Me more than the fact
That degenerate cycles bound.

Ralph's fellowship was to continue for another year; I was due to return to Cambridge, and persuaded him to join me there instead of staying in Princeton. People were not thrown together as closely in Cambridge as in Fine Hall and the Graduate College. Mathematics had no department building then, so mathematicians tended to meet in Hardy's "conversation class" (= seminar) and in the libraries; nevertheless, Ralph and I contrived to see a good deal of each other.

The talks in the conversation class were of variable quality; a particularly dull one by a visiting speaker from Swansea provoked a couple of clerihews from Ralph:

Mr. Wilson on powers
Goes on for hours and hours;
I think that his theory
Is dreadfully dreary.

And:

Is there anything lowlier
Than the singularities of Polya?
By the time they are classified
The audience will be ossified.

In Cambridge Ralph took up candid camera photography; Tukey had introduced us to the art in Princeton. He got some good shots of Hardy, Littlewood, Besicovitch, Heilbronn, etc.; Pat Moran also indulged, and on occasion the snapper was snapped while snapping.

Ralph's family (his parents and his sister Marie, later Marie Boas Hall and well known as a historian of science) spent some time in London in the spring of 1939, and we all met there on several occasions. In the Easter vacation Ralph visited my native city of Edinburgh for a few days, and I was able to show him some of the local sights.

The climax of that academic year, as far as we were concerned, came in the Easter term. André Weil, Claude Chabauty, and Louis Bouckaert (from Louvain) were all in Cambridge, and the proposal was mooted that a marriage should be arranged between Bourbaki's daughter Betti and Hector Pétard; the marriage announcement was duly printed in the canonical French style—on it Pétard was described as the ward of Ersatz Stanislas Pondiczery—and it was circulated to the friends of both parties. A couple of weeks later the Weils, Louis Bouckaert, Max Krook (a South African astrophysicist), Ralph, and myself made a river excursion to Grantchester

Monsieur Nicolas Bourbaki, Membre Canonique de l'Académie Royale de Poldévie, Grand Maître de l'Ordre des Compacts, Conservateur des Uniformes, Lord Protecteur des Filtres, et Madame, née Biunivoque, ont l'honneur de vous faire part du mariage de leur fille Betti avec Monsieur Hector Pétard, Administrateur-Délégué de la Société des Structures Induites, Membre Diplômé de l'Institute of Class-field Archaeologists, Secrétaire de l'Œuvre du Sou du Lion.

Monsieur Ersatz Stanislas Pondiczery, Complexe de Recouvrement de Première Classe en retraite, Président du Home de Rééducation des Faiblement Convergents, Chevalier des Quatre U, Grand Opérateur du Groupe Hyperbolique, Knight of the Total Order of the Golden Mean, L.U.B., C.C., H.L.C., et Madame, née Compactensoi, ont l'honneur de vous faire part du mariage de leur pupille Hector Pétard avec Mademoiselle Betti Bourbaki, ancienne élève des Bienordonnées de Besse.

L'isomorphisme trivial leur sera donné par le P. Adique, de l'Ordre des Diophantiens, en la Cohomologie principale de la Variété Universelle, le 3 Cartembre, an VI, à l'heure habituelle.

L'orgue sera tenu par Monsieur Modulo, Assistant Simplexe de la Grassmannienne (lemmes chantés par la Scholia Cartanorum). Le produit de la quête sera versé intégralement à la maison de retraite des Pauvres Abstraits. La convergence sera assurée.

Après la congruence, Monsieur et Madame Bourbaki recevront dans leurs domaines fondamentaux. Sauterie avec le concours de la fanfare du 7e Corps Quotient.

Tenue canonique
(idéaux à gauche à la boutonnière). C. Q. F. D.

The invitation to Betti Bourbaki's wedding.

Pétard has captured the lion. From left to right: Boas, Smithies, Weil.
Grantchester, 13 May 1939.

by punt and canoe to have tea at the Red Lion; there is a photograph (of which I still have a copy) of Ralph and myself, with our triumphantly captured lion between us, and André Weil looking benevolently on; from the same occasion derives a picture of Weil looking coyly over the top of some of the first proof-sheets of Bourbaki.

A. Weil with proofs of the first number of Bourbaki, Grantchester, 13 May 1939.

Wooden highlander outside tobacconist's shop, Norwich, 12 June 1939.

Later in that term, in the dead period after the end of lectures, when the students were undergoing examinations, Ralph and I made several day excursions: to Norwich, where we admired the then new City Hall, and also a wooden Highlander outside a tobacconist's shop, playing the same role as a wooden Indian in America; to Whipsnade, where Ralph insisted on seeing the wombat, because he liked its name; and to London, where Reginald Sorensen, M.P., got us tickets to attend some of a debate in the House of Commons, a dull affair, enlivened only by the Minister of Agriculture's ignorance about bere barley, a special variety grown in the far north of Scotland.

Ralph stayed on in Cambridge for a while after I left for Scotland; he went down with appendicitis, was operated on, and took advantage of his stay in hospital to grow an impressive beard (of which I have a photograph).

He arrived back in the States just as World War II was breaking out. We began an intensive correspondence; in the first year of the war we must have exchanged more than 60 letters and postcards. The correspondence continued with gradually diminishing intensity; my RPB file is several inches thick. We exchanged offprints, and vast quantities of gossip, mainly mathematical; we delved into numerous oddities, a recent one being how the curve known as the witch of Agnesi came to be called the "lokon" in Russian (a still unsolved problem; we thought we had a clue, but it led nowhere). We exchanged opinions about world affairs; I have always remembered

R. P. Boas with beard, London, August 1939.

his furious reaction to some extremist statement (I have forgotten who made it): "anyone who would die for a cause ought to be shot."

Ralph never came to Europe again, but I visited him in Cambridge (Mass.) in 1950, in Evanston in 1965 and 1981, and in Seattle in May 1992. On that occasion we exchanged gossip and opinions for a couple of hours.

By a curious coincidence, the news of his death came to me from his sister Marie, whom I encountered at a party in Cambridge to celebrate the publication of a Festschrift for D. T. Whiteside's 60th birthday; we had not actually met since 1939, though I had attended one or two lectures that she had given in Cambridge on the history of science.

REMARKS: MEMORIAL FOR RALPH P. BOAS 9 OCTOBER 1992

Harold P. Boas

This is an approximation, reconstructed from memory, to what I said at the memorial—or to what I would have said if I had written it down ahead of time.

The bow tie that I am wearing is one of my father's, made, as my sister mentioned, by my mother.

I hope those of you in the audience will not be offended if I think of you as members of a calculus class.

This is an occasion on which we have gathered together to share memories. It occurred to me that on such an occasion my father would probably have told stories, so I would like to tell you some stories. Some of them are my father's, and some of them are my own.

Let me begin with a story about Zygmund, for some of you may have been just yesterday to the memorial for Zygmund. In my father's early days at Northwestern, he used to commute into Chicago to attend Zygmund's famous seminar. He told me once how it worked. Each week, whoever had something interesting to say would get up and speak. If nobody else had anything interesting to talk about, then Zygmund would speak—and Zygmund always had something interesting to talk about. I have been told that, in the days when I was still of a size to be carried about in a basket, such commuting sometimes resulted in my parents meeting at an El station to hand me off from one to the other.

After my father became department chairman and was busy raising a family, he stopped going to Zygmund's seminar. Now I met Zygmund in 1978 at a conference in Williamstown, Massachusetts, and he said to me, "Your father doesn't come to our seminar any more." It had been perhaps twenty years since my father had attended Zygmund's seminar!

My experiences growing up were somewhat different from my sister's, for as I studied mathematics in college and graduate school, I found that I had tremendous name recognition. When I was introduced to people, I was constantly being asked, "Are you related to... ?" Now sometimes it would be, "Are you related to the author of the book on mathematical methods in the physical sciences?", and I would say, "Yes, that's my mother"; but since I moved in mathematical circles, it was more often, "Are you related to Ralph Boas?" I remember that in high school, when I was prominent on the chess circuit, my father once came home from the university and related rather proudly that he had been introduced to someone on campus, and he had been asked, "Oh, are you related to the chess player?"

When I was at Columbia University, Paul Erdős, the famous eccentric and extraordinarily prolific mathematician, passed through to give a lecture. Someone, probably Lipman Bers, pointed me out to Erdős, and he came over to recite for me the contents of a rhymed telegram that had been sent to my parents on the occasion of their marriage. I am afraid that these verses are lost to posterity, but I remember that they began something like "Ralph and Mary/Now binary/No longer solitary" and continued on in a pseudo-mathematical vein.* This impressed me more than Erdős's hundreds of papers, that he had at the tip of his tongue the contents of a forty-year-old telegram.

Many of you in the audience have heard of the so-called Erdős number, which is the length of the shortest chain of co-authors connecting one to Erdős. Thus, my father's Erdős number is one, because he wrote a joint paper with Erdős. My Erdős number is two, because I wrote a paper with my father, who wrote a paper with Erdős. My collaborators have Erdős number three, unless they are connected to Erdős by some other shorter chain of co-authors. My father used to claim that Kaplansky has Erdős number zero, because Kaplansky wrote one of Erdős's papers!

This story must not be taken too literally, because the principals remember it in different ways, but since my father's version is too good to be lost from oral tradition, I would like to tell it to you now.† The way my father used to tell it is the following. Erdős wrote a paper (with a collaborator) which proved, among other things, the now well-known theorem that if you have an infinite number of points in the plane such that the distance between each pair of points is always an integer, then

---

*Note added February 1993: I have since learned that the verses were "Here's to Boas, Ralph and Mary,/Who no longer solitary/Constitute a form binary./This occasion celebrary/Brings this wire felicitary/From the house of Pondiczery."

†Note added January 1993: A version different from my father's memory of events has appeared in the *Mathematical Intelligencer* 14 (1992), no. 1, 56–57.

Ralph P. Boas Jr., in the club jacket of the Friends of Pondiczery.

the points must actually lie on a straight line.§ This was in the days when my father was running *Mathematical Reviews*, and he sent the paper to Kaplansky to review. While reviewing the paper, Kaplansky discovered the now standard one-paragraph proof of the theorem. Kaplansky thought, "Erdős would like this", but Erdős was off in Timbuctoo, or East Poldavia, or somewhere, so Kaplansky wrote up the proof, signed Erdős's name to it, and sent it off to be published.

In the course of time, the paper was published and came to *Mathematical Reviews* to be reviewed. My father, with a strong sense of symmetry, wrote the review himself and signed Kaplansky's name to it.‡

In 1938, a famous tongue-in-cheek paper on the mathematical theory of big game hunting was published under the name of H. Pétard in the *American Mathematical Monthly*. Although it was once a closely held secret, it is generally known now that the co-conspirators on this paper were my father and Frank Smithies. It is perhaps not generally known that Pétard's full initials are H. W. O., standing for "Hoist

§Note added July 1993: This was subsequently set as a problem in the eighteenth William Lowell Putnam Mathematical Competition, February, 1958.

‡The review simply quotes the paper in its entirety. See *Math. Rev.* 7 (1946), p. 164.

With Own." Actually H. Pétard is the pen-name for the imaginary mathematician E. S. Pondiczery, and is thus a second-order pseudonym.

To tell you about E. S. Pondiczery, I have to change my costume.

[takes off coat and picks up green coat]

My father had this jacket made for him in England. It has a label inside the collar that reads "R. P. Boas, July 1939," and as you can see, it fits me rather well.

[puts on green jacket]

This coat is supposed to be a sort of club jacket for the Friends of Pondiczery. In mathematics, a capital Greek letter Σ denotes sum, or the totality of members of a set. Thus "Sigma Pondiczery" could indicate the collection of members or friends of Pondiczery. This explains the emblem on the pocket of the jacket, which consists of a cherry with a Greek Σ on top of it: Σ 'pon duh cherry. Pondiczery's initials E. S. were so chosen because in the original conception, E. S. Pondiczery was going to write spoof papers on Extra-Sensory Perception.

I would like to close by quoting from a letter I received last month from Frank Smithies in England. He captured the way many of us remember my father. Frank Smithies wrote of my father, "He had an inexhaustible interest in the little oddities of mathematics and of life in general, and when one of them is encountered I shall always feel that I wanted to tell him about it."

December 1992

MY MEMORIES OF RALPH BOAS

Deborah Tepper Haimo
Past President, Mathematical Association of America
University of Missouri, St. Louis

Ralph Boas was a major figure in mathematics, an outstanding educator, and above all, an exceptional human being. In his professional life, Ralph contributed in many capacities and on many fronts, but his contributions to the MAA and his loyalty and devotion to the organization were extraordinary.

Throughout his life, Ralph never refused an assignment if health was not a problem and he always had much to offer. He held many MAA posts at all levels, including a term as President when the effective, calm manner in which he presided at its national meetings provided a most impressive model.

Ralph served as Editor of the *American Mathematical Monthly*, and when he was invited to take part in the *Monthly* Centennial Celebration to be held in San Antonio in January of 1993, he wrote back: "I would love to participate in the celebration of the hundredth anniversary of the *Monthly*, but it's a physical impossibility, ... I can still do useful mental work, but I have to get around the house with an electric cart."

His mind remained keen until the end when he was still working hard on a new edition of his Carus Monograph, *A Primer of Real Functions*. How meritorious Ralph was of our most prestigious honor, the MAA Award for Distinguished Service to Mathematics, which, about a decade earlier, we had conferred on him!

Ralph left an indelible mark on those with whom he came in contact. He had a major impact on my own life in some very personal experiences. As young graduate students, my late husband, Franklin, and I met Ralph at Harvard in the mid '40s. When we were invited to the Boas's house to dinner, he and Mary had prepared everything together, sharing all tasks in a way that was almost unheard of in those days, though generally common now in similar circles. Since both Frank and I came from traditional backgrounds where a man was never seen in the kitchen, it was a major revelation, which completely transformed our lives.

When Ralph assumed his post as editor of *Mathematical Reviews*, he needed someone to review Russian papers and asked Frank for help, thus adding the study of a new language to Frank's long-standing interest in foreign tongues.

It was Ralph, too, who later cleared the way for my becoming a member of the MAA in my own right. No special provisions existed then to encourage independent membership by two professional members of one family, and, as MAA President, he wanted to be able to appoint me to a committee.

Ralph touched many lives in very many ways. He will always live in our hearts and memories.

OBITUARY*

Philip J. Davis
Brown University

He was my thesis advisor. It came about in this way. When I returned to graduate school at the conclusion of World War II, I used to go to the Harvard-MIT mathematical colloquium fairly regularly. At that time Ralph was managing editor of the *Mathematical Reviews*, with editorial offices on the Brown campus in Providence, but he lived in Cambridge. He also attended this colloquium fairly regularly, and we used to walk home together afterward. From these walks and talks there ensued a friendship. He suggested a thesis topic within the theory of infinite interpolation for entire functions of exponential type.

The Department of Mathematics at Harvard was quite willing for me to work with him. Of course, as a young graduate student, I was thrilled to be apprenticed to someone so brilliant, and the fact that Ralph had sat at the feet of G. H. Hardy and Norbert Wiener added an extra cachet, an extra bit of romance, to my position. We all like to trace our mathematical descent back through the Great Names. His suggestions led to my thesis in 1950. At the time, the theory of entire functions was his great love, and in 1954 he published *Entire Functions*, a book that has remained a definitive text on the subject.

In appearance, dapper—always bow-tied, with his glasses and moustache he reminded me fondly of a refined version of Groucho Marx. He was slight and agile; he jumped up on desks with balletic flair; and in the days when he was a member of the Otto Neugebauer group at Brown University (Neugebauer had founded the *Mathematical Reviews*), he was nicknamed "The Squirrel." Many remember Ralph walking down Garden Street, green bookbag over his shoulder, on the way to Harvard Square, South Station, and Providence through sun, rain, and snow.

---

*Reprinted with permission from *SIAM News*, Vol. 26, January 1993.

By 1950 he had left Cambridge to take a position at Northwestern University, and several years later, I went to the National Bureau of Standards in Washington. In the decades that followed, we saw one another but rarely, mostly at national gatherings. For several summers when our children were small, we used to meet on the Cape Cod beaches near the summer house in Orleans his parents had bought in the '30s.

Ralph Boas grew up in an academic environment. His father and mother were both professors of English at Wheaton College in Norton, Massachusetts. His humanistic inheritance from his parents led to a knowledge and a sophistication in science and the humanities that were both wide and deep. His sense of mathematical aesthetics was sharp. Throughout his life he wrote poetry. Perhaps one day a collection will be published.

In addition to *Entire Functions*, Boas wrote *Integrability Theorems for Trigonometric Transforms* (1967) and the very popular and widely studied *A Primer of Real Functions* (1960). A collaboration with Creighton Buck resulted in *Polynomial Expansions of Analytic Functions* (1958).

Boas edited and translated numerous foreign texts. He worked on Russian-English mathematical dictionaries, and, for a number of years, was editor of *The American Mathematical Monthly*. He was a past president of the Mathematical Association of America. All of this displays a tremendous effort on behalf of public mathematics and the infrastructure of our subject, work that is unsung and under-rewarded but without which the whole enterprise crashes.

A list of publications, of positions, of students, of honors, is rarely an adequate summary of an intellectual life. We are all born into a world, including a mathematical world, that we did not make. This world nurtures us and expects something in return. To some, the only possible return consists of a bundle of technical papers. Others see the necessity also of increasing the comprehensibility, the availability, the utility of what the next person has done. Ralph Boas's sense of obligation to the World Mathematical Community was strong. We profit from his labor, and as we treasure it as an inheritance, we should allow ourselves in some measure to walk along his path.

SECTION 1

LION HUNTING

As explained in the reminiscence of Frank Smithies earlier and in the autobiographical essay by Boas himself, the collection of methods for catching a lion that they published under the pseudonym, H. Pétard, appeared in the American Mathematical Monthly in 1938. As is evident from the other articles in this section, the idea prompted a good many others to add to this literature. We include those articles of which we are aware—we make no claim that this is a complete compendium of contributions to this area of mathematics.

A CONTRIBUTION TO THE MATHEMATICAL THEORY OF BIG GAME HUNTING*

H. Pétard
Princeton, New Jersey

This little known mathematical discipline has not, of recent years, received in the literature the attention which, in our opinion, it deserves. In the present paper we present some algorithms which, it is hoped, may be of interest to other workers in the field. Neglecting the more obviously trivial methods, we shall confine our attention to those which involve significant applications of ideas familiar to mathematicians and physicists.

The present time is particularly fitting for the preparation of an account of the subject, since recent advances both in pure mathematics and in theoretical physics have made available powerful tools whose very existence was unsuspected by earlier investigators. At the same time, some of the more elegant classical methods acquire new significance in the light of modern discoveries. Like many other branches of knowledge to which mathematical techniques have been applied in recent years, the Mathematical Theory of Big Game Hunting has a singularly happy unifying effect on the most diverse branches of the exact sciences.

For the sake of simplicity of statement, we shall confine our attention to Lions (*Felis leo*) whose habitat is the Sahara Desert. The methods which we shall enumerate will easily be seen to be applicable, with obvious formal modifications, to other carnivores and to other portions of the globe. The paper is divided into three parts, which draw their material respectively from mathematics, theoretical physics, and experimental physics.

---

*_Amer. Math. Monthly_ 45 (1938), 446–447; reprinted with additions, _Eureka_ (Cambridge, England), 1939.

The author desires to acknowledge his indebtedness to the Trivial Club of St. John's College, Cambridge, England; to the M.I.T. chapter of the Society for Useless Research; to the F. o. P., of Princeton University; and to numerous individual contributors, known and unknown, conscious and unconscious.

1. MATHEMATICAL METHODS

1. The Hilbert, or Axiomatic, Method. We place a locked cage at a given point of the desert. We then introduce the following logical system.

Axiom I. *The class of lions in the Sahara Desert is non-void.*

Axiom II. *If there is a lion in the Sahara Desert, there is a lion in the cage.*

Rule of Procedure. *If p is a theorem, and "p implies q" is a theorem, then q is a theorem.*

Theorem 1. *There is a lion in the cage.*

2. The Method of Inversive Geometry. We place a *spherical* cage in the desert, enter it, and lock it. We perform an inversion with respect to the cage. The lion is then in the interior of the cage, and we are outside.

3. The Method of Projective Geometry. Without loss of generality, we may regard the Sahara Desert as a plane. Project the plane into a line, and then project the line into an interior point of the cage. The lion is projected into the same point.

4. The Bolzano-Weierstrass Method. Bisect the desert by a line running N–S. The lion is either in the E portion or the W portion; let us suppose him to be in the W portion. Bisect this portion by a line running E–W. The lion is either in the N portion or in the S portion; let us suppose him to be in the N portion. We continue this process indefinitely, constructing a sufficiently strong fence about the chosen portion at each step. The diameter of the chosen portions approaches zero, so that the lion is ultimately surrounded by a fence of arbitrarily small perimeter.

5. The "Mengentheoretisch" Method. We observe that the desert is a separable space. It therefore contains an enumerable dense set of points, from which can be extracted a sequence having the lion as limit. We then approach the lion stealthily along this sequence, bearing with us suitable equipment.

6. The Peano Method. Construct, by standard methods, a continuous curve passing through every point of the desert. It has been remarked* that it is possible to traverse such a curve in an arbitrarily short time. Armed with a spear, we traverse the curve in a time shorter than that in which a lion can move his own length.

---

*By Hilbert. See E. W. Hobson, The Theory of Functions of a Real Variable and the Theory of Fourier's Series, 1927, vol. 1, pp. 456–457.

7. A TOPOLOGICAL METHOD. We observe that a lion has at least the connectivity of the torus. We transport the desert into four-space. It is then possible** to carry out such a deformation that the lion can be returned to three-space in a knotted condition. He is then helpless.

8. THE CAUCHY, OR FUNCTION THEORETICAL, METHOD. We consider an analytic lion-valued function $f(z)$. Let ζ be the cage. Consider the integral

$$\frac{1}{2\pi i}\int_C \frac{f(z)}{z-\zeta}\,dz,$$

where C is the boundary of the desert; its value is $f(\zeta)$, i.e., a lion in the cage.†

9. THE WIENER TAUBERIAN METHOD. We procure a tame lion, L_0, of class $L(-\infty, \infty)$, whose Fourier transform nowhere vanishes, and release it in the desert. L_0 then converges to our cage. By Wiener's General Tauberian Theorem,‡ any other lion, L (say), will then converge to the same cage. Alternatively, we can approximate arbitrarily closely to L by translating L_0 about the desert.§

2. METHODS FROM THEORETICAL PHYSICS

10. THE DIRAC METHOD. We observe that wild lions are, *ipso facto*, not observable in the Sahara Desert. Consequently, if there are any lions in the Sahara, they are tame. The capture of a tame lion may be left as an exercise for the reader.

11. THE SCHRÖDINGER METHOD. At any given moment there is a positive probability that there is a lion in the cage. Sit down and wait.

12. THE METHOD OF NUCLEAR PHYSICS. Place a tame lion in the cage, and apply a Majorana exchange operator|| between it and a wild lion.

As a variant, let us suppose, to fix ideas, that we require a male lion. We place a tame lioness in the cage, and apply a Heisenberg exchange operator ¶ which exchanges the spins.

**H. Seifert and W. Threlfall, Lehrbuch der Topologie, 1934, pp. 2–3.

†N.B. By Picard's Theorem (W. F. Osgood, Lehrbuch der Funktionentheorie, vol. 1, 1928, p. 748), we can catch every lion with at most one exception.

‡N. Wiener, The Fourier Integral and Certain of its Applications, 1933, pp. 73–74.

§N. Wiener, *l. c.*, p. 89.

||See, for example, H. A. Bethe and R. F. Bacher, Reviews of Modern Physics, vol. 8, 1936, pp. 82–229; especially pp. 106–107.

¶*Ibid.*

13. A RELATIVISTIC METHOD. We distribute about the desert lion bait containing large portions of the Companion of Sirius. When enough bait has been taken, we project a beam of light across the desert. This will bend right round the lion, who will then become so dizzy that he can be approached with impunity.

3. METHODS FROM EXPERIMENTAL PHYSICS

14. THE THERMODYNAMICAL METHOD. We construct a semi-permeable membrane, permeable to everything except lions, and sweep it across the desert.

15. THE ATOM-SPLITTING METHOD. We irradiate the desert with slow neutrons. The lion becomes radioactive, and a process of disintegration sets in. When the decay has proceeded sufficiently far, he will become incapable of showing fight.

16. THE MAGNETO-OPTICAL METHOD. We plant a large lenticular bed of catnip (*Nepeta cataria*), whose axis lies along the direction of the horizontal component of the earth's magnetic field, and place a cage at one of its foci. We distribute over the desert large quantities of magnetized spinach (*Spinacia oleracea*), which, as is well known, has a high ferric content. The spinach is eaten by the herbivorous denizens of the desert, which are in turn eaten by lions. The lions are then oriented parallel to the earth's magnetic field, and the resulting beam of lions is focussed by the catnip upon the cage.

A NEW METHOD OF CATCHING A LION*

I.J. Good

In this note a definitive procedure will be provided for catching a lion in a desert (see [1]).

Let Q be the operator that encloses a word in quotation marks. Its square Q^2 encloses a word in double quotes. The operator clearly satisfies the law of indices, $Q^m Q^n = Q^{m+n}$. Write down the word 'lion,' without quotation marks. Apply to it the operator Q^{-1}. Then a lion will appear on the page. It is advisable to enclose the page in a cage before applying the operator.

[1] H. Pétard, A contribution to the mathematical theory of big game hunting, this *Monthly* 45 (1938) 446–447.

**Amer. Math. Monthly* 72 (1965), 436.

ON A THEOREM OF H. PÉTARD*

Christian Roselius
Tulane University

In a classical paper [4], H. Pétard proved that it is possible to capture a lion in the Sahara desert. He further showed [4, no. 8, footnote] that it is in fact possible to catch every lion with at most one exception. Using completely new techniques, not available to Pétard at the time, we are able to sharpen this result, and to show that *every* lion may be captured.

Let $\mathcal{L}$ denote the category whose objects are lions, with "ancestor" as the only nontrivial morphism. Let ℓ be the category of caged lions. The subcategory ℓ is clearly complete, is nonempty (by inspection), and has both a generator and cogenerator [3, vii, 15–16]. Let $F : \ell \to \mathcal{L}$ be the forgetful functor, which forgets the cage. By the Adjoint Functor Theorem [1, 80–91] the functor F has a coadjoint $C : \mathcal{L} \to \ell$, which reflects each lion into a cage.

We remark that this method is obviously superior to the Good method [2], which only guarantees the capture of one lion, and which requires an application of the Weierkäfig Preparation Theorem.

[1] P. Freyd, Abelian categories, New York, 1964.

[2] I. J. Good, A new method of catching a lion, this *Monthly*, 72 (1965) 436.

[3] Moses, The Book of Genesis.

[4] H. Pétard, A contribution to the mathematical theory of big game hunting, this *Monthly*, 45 (1938) 446–447.

**Amer. Math. Monthly* 74 (1967), 838–839.

SOME MODERN MATHEMATICAL METHODS IN THE THEORY OF LION HUNTING*

Otto Morphy, D.Hp.
(Dr. of Hypocrisy)

It is now 30 years since the appearance of H. Pétard's classic treatise [2] on the mathematical theory of big game hunting. These years have seen a remarkable development of practical mathematical techniques. It is, of course, generally known that it was Pétard's famous letter to the president in 1941 that led to the establishment of the Martini Project, the legendary crash program to develop new and more efficient methods for search and destroy operations against the axis lions. The Infernal Bureaucratic Federation (IBF) has recently declassified certain portions of the formerly top secret Martini Project work. Thus we are now able to reveal to the world, for the first time, these important new applications of modern mathematics to the theory and practice of lion hunting. As has become standard practice in the discipline [2] we shall restrict our attention to the case of lions residing in the Sahara Desert [3]. As noted by Pétard, most methods apply, more generally, to other big game. However, method (3) below appears to be restricted to the genus Felis. Clearly, more research on this important matter is called for.

1. Surgical Method. A lion may be regarded as an orientable three-manifold with a nonempty boundary. It is known [4] that by means of a sequence of surgical operations (known as "spherical modifications" in medical parlance) the lion can be rendered contractible. He may then be signed to a contract with Barnum and Bailey.

**Amer. Math. Monthly* 75 (1968), 185–187.

2. LOGICAL METHOD. A lion is a continuum. According to Cohen's theorem [5] he is undecidable (especially when he must make choices). Let two men approach him simultaneously. The lion, unable to decide upon which man to attack, is then easily captured.

3. FUNCTORIAL METHOD. A lion is not dangerous unless he is somewhat gory. Thus the lion is a category. If he is a small category then he is a kittygory [6] and certainly not to be feared. Thus we may assume, without loss of generality, that he is a proper class. But then he is not a member of the universe and is certainly not of any concern to us.

4. METHOD OF DIFFERENTIAL TOPOLOGY. The lion is a three-manifold embedded in euclidean 3-space. This implies that he is a handlebody [7]. However, a lion which can be handled is tame and will enter the cage upon request.

5. SHEAF THEORETIC METHOD. The lion is a cross-section [8] of the sheaf of germs of lions [9] on the Sahara Desert. Merely alter the topology of the Sahara, making it discrete. The stalks of the sheaf will then fall apart releasing the germs which attack the lion and kill it.

6. METHOD OF TRANSFORMATION GROUPS. Regard the lion as a surface. Represent each point of the lion as a coset of the group of homeomorphisms of the lion modulo the isotropy group of the nose (considered as a point) [10]. This represents the lion as a homogeneous space. That is, this representation homogenizes the lion. A homogenized lion is in no shape to put up a fight [11].

7. POSTNIKOV METHOD. A male lion is quite hairy [12] and may be regarded as being made up of fibers. Thus we may regard the lion as a fiber space. We may then construct a Postnikov decomposition [13] of the lion. This being done, the lion, being decomposed, is dead and in bad need of burial.

8. STEENROD ALGEBRA METHOD. Consider the mod p cohomology ring of the lion. We may regard this as a module over the mod p Steenrod algebra. Doing this requires the use of the table of Steenrod cohomology operations [14]. Every element must be killed by some of these operations. Thus the lion will die on the operating table.

9. HOMOTOPY METHOD. The lion has the homotopy type of a one-dimensional complex and hence he is a $K(\pi, 1)$ space. If π is noncommutative then the lion is not a member of the international commutist conspiracy [15] and hence he must be friendly. If π is commutative then the lion has the homotopy type of the space of loops on a $K(\pi, 2)$ space [13]. We hire a stunt pilot to loop the loops, thereby hopelessly entangling the lion and rendering him helpless.

10. COVERING SPACE METHOD. Cover the lion by his simply connected covering space. In effect this decks the lion [16]. Grab him while he is down.

11. Game Theoretic Method. A lion is big game. Thus, *a fortiori*, he is a game. Therefore there exists an optimal strategy [17]. Follow it.

12. Group Theoretic Method. If there are an even number of lions in the Sahara Desert we add a tame lion. Thus we may assume that the group of Sahara lions is of odd order. This renders the situation capable of solution according to the work of Thompson and Feit [18].

We conclude with one significant nonmathematical method:

13. Biological Method. Obtain a number of planarians and subject them to repeated recorded statements saying: "You are a planarian." The worms should shortly learn this fact since they must have some suspicions to this effect to start with. Now feed the worms to the lion in question. The knowledge of the planarians is then transferred to the lion [19]. The lion, now thinking that he is a planarian, will proceed to subdivide. This process, while natural for the planarian, is disastrous to the lion [20].

Ed. Note: Prof. Morphy is the namesake of his renowned aunt, the author of the famous series of epigrams now popularly known as Auntie Otto Morphisms or euphemistically as epimorphisms.

Footprints

[1] This report was supported by grant #007 from Project Leo of the War on Puberty.

[2] H. Pétard. A contribution to the mathematical theory of big game hunting, this *Monthly*, (1938).

[3] This restriction of the habitat does not affect the generality of the results because of Brouwer's theorem on the invariance of domain.

[4] Kervaire and Milnor, Groups of homotopy spheres, I, Ann. of Math., (1963).

[5] P. J. Cohen, The independence of the continuum hypothesis, Proc. N.A.S. (63–64).

[6] P. Freyd, Abelian Categories, Harper and Row, New York, 1964.

[7] S. Smale, A survey of some recent developments in differential topology, Bull. A.M.S., (1963).

[8] It has been experimentally verified that lions are cross.

[9] G. Bredon, Sheaf Theory, McGraw-Hill, New York, 1967.

[10] Montgomery and Zippin, Topological Transformation Groups, Interscience, 1955.

[11] E. Borden, Characteristic classes of bovine spaces, Peripherblatt für Math., (1966BC).

[12] Eddy Courant, Sinking of the Mane, Pantz Press, 1898.

[13] E. Spanier, Algebraic Topology, McGraw-Hill, New York, 1966.

[14] Steenrod and Epstein, Cohomology Operations, Princeton, 1962.

[15] Logistics of the Attorney General's list, Band Corp. (1776).

[16] Admiral, T. V., (USN Ret.), How to deck a swab, ONR tech. rep. (classified).

[17] von Neumann and Morgenstern, Theory of Games . . . , Princeton, 1947.

[18] Feit and Thompson, Solvability of groups of odd order, Pac. J. M. (1963).

[19] J. V. McConnell, ed., The Worm Re-turns, Prentice-Hall, Englewood Cliffs, N.J., 1965.

[20] This method must be carried out with extreme caution for, if the lion is large enough to approach critical mass, this fissioning of the lion may produce a violent reaction.

FURTHER TECHNIQUES IN THE THEORY OF BIG GAME HUNTING*

Patricia L. Dudley
G. T. Evans
K. D. Hansen
I. D. Richardson
Carleton University, Ottawa

Interest in the problem of big game hunting has recently been reawakened by Morphy's paper in this *Monthly*, Feb. 1968, p. 185. We outline below several new techniques, including one from the humanities. We are also in possession of a solution by means of Bachmann geometry which we shall be glad to communicate to anyone who is interested.

1. MOORE-SMITH METHOD. Letting A = Saraha Desert, one can construct a net in A converging to any point in $\bar{A}$. Now lions are unable to resist tuna fish, on account of the charged atoms found therein (see Galileo Galilei, *Dialogues Concerning Tuna's Ionses*). Place a tuna fish in a tavern, thus attracting a lion. As noted above, one can construct a net converging to any point in a bar; in this net enmesh the lion.

2. METHOD OF ANALYTICAL MECHANICS. Since the lion has nonzero mass it has moments of inertia. Grab it during one of them.

3. MITTAG-LEFFLER METHOD. The number of lions in the Sahara Desert is finite, so the collection of such lions has no cluster point. Use Mittag-Leffler's theorem to construct a meromorphic function with a pole at each lion. Being a tropical animal a lion will freeze if placed at a pole, and may then be easily taken.

4. METHOD OF NATURAL FUNCTIONS. The lion, having spent his life under the Sahara sun, will surely have a tan. Induce him to lie on his back; he can then, by virtue of his reciprocal tan, be cot.

**Amer. Math. Monthly* 75 (1968), 896–897.

5. Boundary Value Method. As Dr. Morphy has pointed out, Brouwer's theorem on the invariance of domain makes the location of the hunt irrelevant. The present method is designed for use in North America. Assemble the requisite equipment in Kentucky, and await inclement weather. Catching the lion then readily becomes a Storm-Louisville problem.

6. Method of Moral Philosophy. Construct a corral in the Sahara and wait until autumn. At that time the corral will contain a large number of lions, for it is well known that a pride cometh before the fall.

15 NEW WAYS TO CATCH A LION*

John Barrington†

This, O Best Beloved, is another tale of the High and the Far-Off Times. In the blistering midst of the Sand-Swept Sahara lived a Pride of Lions. There was a Real Lion, and a Projective Lion, and a pair of Parallel Lions; and all manner of Lion Segments. And on the edge of the Sand-Swept Sahara there lived a 'nexorable Lion Hunter. . .

I make no apologies for raising once again the problems of the mathematical theory of big game hunting. As with any branch of mathematics, much progress has been made in the last decade.

The subject started in 1938 with the epic paper of Pétard [1]. The main problem is usually formulated as follows: *In the Sahara desert there exist lions. Devise methods for capturing them.* Pétard found ten mathematical solutions, which we can paraphrase as follows.

1. THE HILBERT METHOD. Place a locked cage in the desert. Set up the following axiomatic system.

(i) The set of lions is non-empty.

(ii) If there is a lion in the desert, then there is a lion in the cage.

THEOREM 1. There is a lion in the cage.

2. THE METHOD OF INVERSIVE GEOMETRY. Place a locked, spherical cage in the desert, empty of lions, and enter it. Invert with respect to the cage. This maps the lion to the interior of the cage, and you outside it.

**Seven Years of Manifold/1968–1980*, Ian Stewart and John Jaworski, eds., Cheshire, England, Shiva Publishing Limited, 1981, pp. 36–39. Reprinted by permission.

†Pseudonym for Ian Stewart.

3. THE PROJECTIVE GEOMETRY METHOD. The desert is a plane. Project this to a line, then project the line to a point inside the cage. The lion goes to the same point.

4. THE BOLZANO-WEIERSTRASS METHOD. Bisect the desert by a line running N-S. The lion is in one half. Bisect this half by a line running E-W. The lion is in one half. Continue the process indefinitely, at each stage building a fence. The lion is enclosed by a fence of arbitrarily small length.

5. THE GENERAL TOPOLOGY METHOD. Observe that the desert is a separable metric space, so has a countable dense subset. Some sequence converges to the lion. Approach stealthily along it, bearing suitable equipment.

6. THE PEANO METHOD. There exists a space-filling curve passing through every point of the desert. It has been remarked [2] that such a curve may be traversed in as short a time as we please. Armed with a spear, traverse the curve faster than the lion can move his own length.

7. A TOPOLOGICAL METHOD. The lion has at least the connectivity of a torus. Transport the desert into 4-space. It can now be deformed in such a way as to knot the lion [3]. He is now helpless.

8. THE CAUCHY METHOD. Let $f(z)$ be an analytic lion-valued function, with ζ the cage. Consider the integral

$$\frac{1}{2\pi i}\int_C \frac{f(z)}{z-\zeta}\,dz$$

where C is the boundary of the desert. Its value is $f(\zeta)$, that is, a lion in a cage.

9. THE WIENER TAUBERIAN METHOD. Procure a tame lion L_0 of class $L(-\infty,\infty)$ whose Fourier transform [Furrier transform?] nowhere vanishes, and set it loose in the desert. Being tame, it will converge to the cage. By Wiener [4] every other lion will converge to the same cage.

10. THE ERATOSTHENIAN METHOD.* Enumerate all objects in the desert; examine them one by one; discard all those that are not lions. A refinement will capture only prime lions.

Pétard also gives one physical method with strong mathematical content:

11. THE SCHRÖDINGER METHOD. At any instant there is a non-zero probability that a lion is in the cage. Wait.

The next work of any significance is that of Morphy [5]. I confess that I do not find all of his methods convincing. The best are:

*This differs from the *Monthly* version since it refers to a slightly expanded version of the Pétard article that appeared in *Eureka*.

12. SURGERY. The lion is an orientable 3-manifold with boundary and so [6] may be rendered contractible by surgery. Contract him to Barnum and Bailey.

13. THE COBORDISM METHOD. For the same reasons the lion is a handlebody. A lion that can be handled is trivial to capture.

14. THE SHEAF-THEORETIC METHOD. The lion is a cross-section [8] of the sheaf of germs of lions in the desert. Re-topologize the desert to make it discrete: the stalks of the sheaf fall apart and release the germs, which kill the lion.

15. THE POSTNIKOV METHOD. The lion, being hairy, may be regarded as a fibre space. Construct a Postnikov decomposition [9]. A decomposed lion must, of course, be long dead.

16. THE UNIVERSAL COVERING. Cover the lion by his simply-connected covering space. Since this has no holes, he is trapped!

17. THE GAME-THEORY METHOD. The lion is big game, hence certainly a game. There exists an optimal strategy. Follow it.

18. THE FEIT-THOMPSON METHOD. If necessary add a lion to make the total odd. This renders the problem soluble [10].

Recent, hitherto unpublished, work has revealed a range of new methods:

19. THE FIELD-THEORY METHOD. Irrigate the desert and plant grass so that it becomes a field. A zero lion is trivial to capture, so we may assume the lion $L \neq 0$. The element 1 may be located just to the right of 0 in the prime subfield. Prize it apart into LL^{-1} and discard L^{-1}. (Remark: the Greeks used the convention that the product of two lions is a rectangle, not a lion; the product of 3 lions is a solid, and so on. It follows that every lion is transcendental. Modern mathematics permits algebraic lions.)

20. THE KITTYGORY METHOD. Form the category whose objects are the lions in the desert, with trivial morphisms. This is a small category (even if lions are big cats) and so can be embedded in a concrete category [11]. There is a forgetful functor from this to the category of sets: this sets the concrete and traps the embedded lions.

21. BACKWARD INDUCTION. We prove by backward induction the statement $L(n)$: "It is possible to capture n lions." This is true for sufficiently large n since the lions will be packed like sardines and have no room to escape. But trivially $L(n + 1)$ implies $L(n)$ since, having captured $n + 1$ lions, we can release one. Hence $L(1)$ is true.

22. ANOTHER TOPOLOGICAL METHOD. Give the desert the *leonine* topology, in which a subset is closed if it is the whole desert, or contains no lions. The set of lions is now dense. Put an *open* cage in the desert. By density it contains a lion. Shut it quickly!

23. THE MOORE-SMITH METHOD. Like (5) above, but this applies to non-separable deserts: the lion is caught not by a sequence, but by a net.

24. FOR THOSE WHO INSIST ON SEQUENCES. The real lion is non-compact and so contains non-convergent subsequences. To overcome this let Ω be the first uncountable ordinal and insert a copy of the given lion between α and $\alpha + 1$ for all ordinals $\alpha < \Omega$. You now have a *long lion* in which all sequences converge [12]. Proceed as in (5).

25. THE GROUP RING METHOD. Let Γ be the free group on the set G of lions, and let $Z\Gamma$ be its group ring. The lions now belong to a ring, so are circus lions, hence tame.

26. THE BOURBAKI METHOD. The capture of a lion in a desert is a special case of a far more general problem. Formulate this problem and find necessary and sufficient conditions for its solution. The capture of a lion is now a trivial corollary of the general theory, which *on no account should be written down explicitly.*

27. THE HASSE-MINKOWSKI METHOD. Consider the lion-catching problem modulo p for all primes p. There being only finitely many possibilities, this can be solved. Hence the original problem can be solved [13].

28. THE PL METHOD. The lion is a 3-manifold with non-empty boundary. Triangulate it to get a *PL* manifold. This can be collared [14], which is what we wish to achieve.

29. THE SINGULARITY METHOD. Consider a lion in the plane. If it is a regular lion its regular habits render it easy to catch (e.g. dig a pit). WLOG it is a singular lion. Stable singularities are dense, so WLOG the lion is stable. The singularity is not a self-intersection (since a self-intersecting lion is absurd) so it must be a cusp. Complexify and intersect with a sphere to get a trefoil knot. As in (7) the problem becomes trivial.

30. THE MEASURE-THEORETIC METHOD. Assume for a contradiction that no lion can be captured. Since capturable lions are imaginary, all lions are real. On any real lion there exists a non-trivial invariant measure μ, namely Haar or Lebesgue measure. Then $\mu \times \mu$ is a Baire measure on $L \times L$ by [15]. Since a product of lions cannot be a bear, the Baire measure on $L \times L$ is zero. Hence $\mu = 0$, a contradiction. Thus all lions may be captured.

31. THE METHOD OF PARALLELS. Select a point in the desert and introduce a tame lion not passing through that point. There are three cases:

(a) The geometry is Euclidean. There is then a unique parallel lion passing through the selected point. Grab it as it passes.

(b) The geometry is hyperbolic. The same method will now catch infinitely many lions.

(c) The geometry is elliptic. There are no parallel lions, so every lion meets every other lion. Follow a tame lion and catch all the lions it meets: in this way every lion in the desert will be captured.

32. The Thom-Zeeman Method. A lion loose in the desert is an obvious catastrophe [16]. It has three dimensions of control (2 for position, 1 for time) and one dimension of behaviour (being parametrized by a lion). Hence by Thom's Classification Theorem it is a swallowtail. A lion that has swallowed its tail is in no state to avoid capture.

33. The Australian Method. Lions are very varied creatures, so there is a variety of lions in the desert. This variety contains free lions [17] which satisfy no non-trivial identities. Select a lion and register it as "Fred Lion" at the local Register Office: it now has a non-trivial identity, hence cannot be free. If it is not free it must be captive. (If "Fred Lion" is thought to be a trivial identity, call it "Albert Einstein.")

Bibliography

[1] H. Pétard, *A contribution to the mathematical theory of big game hunting*, Amer. Math. Monthly 45 (1938).

[2] E. W. Hobson, *The theory of functions of a real variable and the theory of Fourier's series*, 1927.

[3] H. Seifert and W. Threlfall, *Lehrbuch der Topologie*, 1934.

[4] N. Wiener, *The Fourier integral and certain of its applications*, 1933.

[5] O. Morphy, *Some modern mathematical methods in the theory of lion hunting*, Amer. Math. Monthly 75 (1968) 185–7.

[6] M. Kervaire and J. Milnor, *Groups of homotopy spheres I*, Ann. of Math. 1963.

[7] This footnote has been censored by the authorities.

[8] It has been verified experimentally that lions are cross.

[9] E. Spanier, *Algebraic Topology*, McGraw-Hill 1966.

[10] W. Feit and J. G. Thompson, *Solvability of groups of odd order*, Pac. J. Math. 1963.

[11] P. Freyd, *Abelian Categories*.

[12] J. L. Kelley, *General Topology*.

[13] J. Milnor and D. Husemoller, *Symmetric Bilinear Forms* 1973.

[14] C. P. Rourke and B. L. Sanderson, *Introduction to Piecewise Linear Topology* 1973.

[15] S. K. Berberian, *Topological Groups*.

[16] R. Thom, *Stabilité Structurelle et Morphogénèse* 1972.

[17] Hanna Neumann, *Varieties of Groups* 1972.

LION-HUNTING WITH LOGIC*†

Houston Euler
c/o Leon Harkleroad
Louisville, KY

Over the years there has developed a body of literature on the use of mathematical techniques to catch lions ([1]–[6]). In this literature there has been a comparative shortage of proofs based on mathematical logic. If those of us who commit logic believe in the vitality of our field, we cannot afford to allow such a shortage to continue. The following proofs, then, are offered as a first step towards rectifying the situation.

1. NONSTANDARD ANALYSIS. In a nonstandard universe (namely, the land of Oz [7]), lions are cowardly and may be caught easily. By the transfer principle, this likewise holds in our (standard) universe.

2. SET THEORY. If the set of lions is bounded, you can simply build a cage around the boundary. So assume that the set of lions is unbounded. It will then have an element in common with a stationary set. But a stationary lion is trivial to capture.

3. SET THEORY. Assume $V = L$. Since the lion is in the universe, it is constructible. So just carry out its construction within a cage in the first place.

4. SET THEORY. Assume AC. Perform a Tarski-Banach decomposition on the lion to halve its size. Repeat until the lion is small enough to be captured easily.

5. RECURSION THEORY. Assume you can capture a lion. Having done so, you can easily bring it to a standstill, and you would thus have a solution to the halting problem. Since the halting problem is unsolvable, you *cannot* capture a lion after all.

---

*This research received compact support from Tombs grant LWH-42755.

†*Amer. Math. Monthly* 92 (1985), 140.

In conjunction with the previous results, we have

COROLLARY. *Mathematics is inconsistent.*

This corollary, besides being of intrinsic interest, also provides solutions to the Riemann Hypothesis, Fermat's Last Theorem, and other questions (besides giving a proof of the Four-Color Theorem that does not require a computer!).

References

[1] H. Pétard, A contribution to the mathematical theory of big game hunting, this *Monthly*, 45 (1938) 446–7.

[2] I. J. Good, A new method of catching a lion, this *Monthly*, 72 (1965) 436.

[3] C. Roselius, On a theorem of H. Pétard, this *Monthly*, 74 (1967) 838–9.

[4] O. Morphy, Some modern mathematical methods in the theory of lion hunting, this *Monthly*, 75 (1968) 185–7.

[5] P. L. Dudley, G. T. Evans, K. D. Hansen, I. D. Richardson, Further techniques in the theory of big game hunting, this *Monthly*, 75 (1968) 896–7.

[6] J. Barrington, 15 new ways to catch a lion, Manifold, Issue 18 (1976), reprinted in Seven Years of Manifold, Shiva Publishing, 1981.

[7] L. F. Baum, The Wonderful Wizard of Oz, 1900.

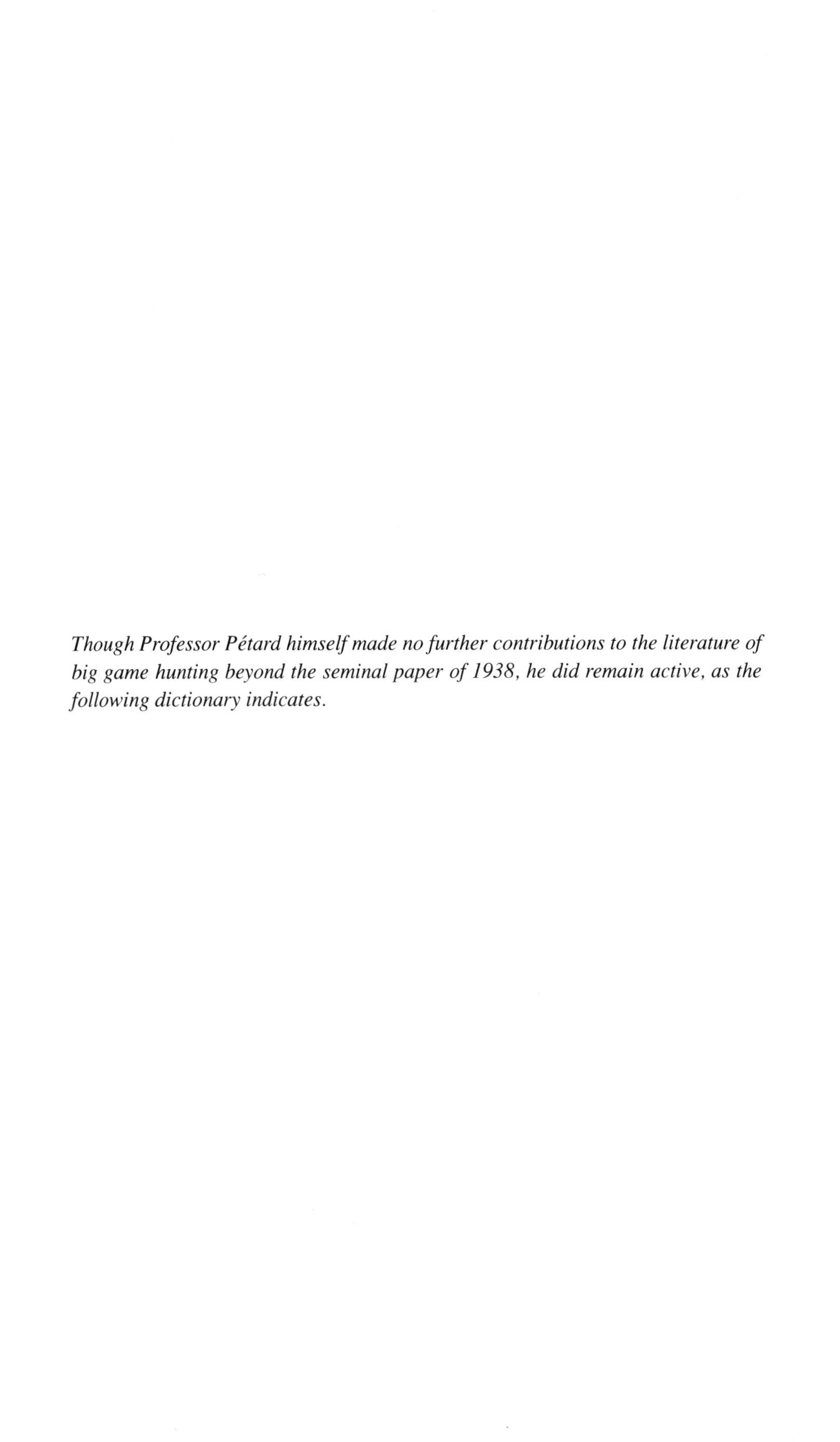

Though Professor Pétard himself made no further contributions to the literature of big game hunting beyond the seminal paper of 1938, he did remain active, as the following dictionary indicates.

A BRIEF DICTIONARY OF PHRASES USED IN MATHEMATICAL WRITING*

H. Pétard
Society for Useless Research

Since authors seldom, if ever, say what they mean, the following glossary is offered to neophytes in mathematical research to help them understand the language that surrounds the formulas. Since mathematical writing, like mathematics, involves many undefined concepts, it seems best to illustrate the usage by interpretation of examples rather than to attempt definition.

ANALOGUE. This is an a. of: I have to have *some* excuse for publishing it.

APPLICATIONS. This is of interest in a.: I have to have *some* excuse for publishing it.

COMPLETE. The proof is now c.: I can't finish it.

DETAILS. I cannot follow the d. of X's proof: It's wrong. We omit the d.: I can't do it.

DIFFICULT. This problem is d.: I don't know the answer. (Cf. Trivial.)

GENERALITY. Without loss of g.: I have done an easy special case.

IDEAS. To fix the i.: To consider the only case I can do.

INGENIOUS. X's proof is i.: I understand it.

INTEREST. It may be of i.: I have to have *some* excuse for publishing it.

*Amer. Math. Monthly 73 (1966), 196–197.

INTERESTING. X's paper is i.: I don't understand it.

KNOWN. This is a k. result but I reproduce the proof for the convenience of the reader: My paper isn't long enough.

Langage. Par abus de l.: In the terminology used by other authors. (Cf. Notation.)

NATURAL. It is n. to begin with the following considerations: We have to start somewhere.

NEW. This was proved by X but the following n. proof may present points of interest: I can't understand X.

NOTATION. To simplify the n.: It is too much trouble to change now.

OBSERVED. It will be o. that: I hope you have not noticed that.

OBVIOUS. It is o.: I can't prove it.

READER. The details may be left to the r.: I can't do it.

REFEREE. I wish to thank the r. for his suggestions: I loused it up.

STRAIGHTFORWARD. By a s. computation: I lost my notes.

TRIVIAL. This problem is t.: I know the answer. (Cf. Difficult.)

WELL-KNOWN. This result is w.: I can't find the reference.

Exercises for the student: Interpret the following.

1. I am indebted to Professor X for stimulating discussions.
2. However, as we have seen.
3. In general.
4. It is easily shown.
5. To be continued.

This article was prepared with the opposition of the National Silence Foundation.

SECTION 2

INFINITE SERIES

CANTILEVERED BOOKS*

It has been pointed out in this Journal[1] that it is possible to stack a pile of identical books on a table so that the top book is as far as you like beyond the edge of the table. The same problem was discussed by Gamow.[2] Having shown that the top book can be made to clear the table if there are as many as 5 books in the stack, Gamow went on to remark that "we will need the entire Library of Congress to make the overhang equal to three or four book lengths." Here, however, Gamow's famous intuition failed him. The amount of overhang attainable with n books is, in book lengths,

$$\frac{1}{2}\left(1 + \frac{1}{2} + \frac{1}{3} + \cdots + n^{-1}\right).$$

It is easily calculated that this exceeds 3 for the first time at $n = 227$ and exceeds 4 for the first time at $n = 1674$, so that a much smaller library will suffice. The harmonic series does diverge very slowly, however, so that a modest increase in the overhang requires many more books; in particular, exactly 272,400,600 would be needed to get an overhang of 10 book lengths, and more than 1.5×10^{44} for a 50 book-length overhang (the exact number is given in Ref. 3).[3]

Amer. J. Phys.* **41 (1973), 715. Reprinted by permission.

[1]L. Eisner, *Amer. J. Phys.* **27**, 121 (1959).

[2]G. Gamow, *Matter, Earth, and Sky* (Prentice-Hall, Englewood Cliffs, N.J., 1958), pp. 15–16.

[3]R. P. Boas, Jr., and J. W. Wrench, Jr., *Amer. Math. Monthly* **78**, 864 (1971).

SNOWFALLS AND ELEPHANTS, POP BOTTLES AND π*†

You must have noticed that we've had some remarkably extreme weather lately. The snowfall in Chicago in the winter of 1977–78, for example, set a record for the years since 1871 (as far back as data exist). The '78–'79 snowfall surpassed even that! Some of us who live near Chicago were surprised. Should we have been? Nobody, after all, is greatly surprised at a new record for the mile run.

Record-breaking weather of any kind is rather rare, in point of fact, whereas athletic records are broken every few days. The reason for the difference, of course, is that athletes are constantly striving to break records; whereas, so far as we know, nobody's actually trying to arrange deeper snows or hotter summers or heavier rains for Chicago. If somebody—for instance, bug-eyed monsters in UFOs—were deliberately modifying our weather, new records ought to turn up much more often than they do.

How many times in your life should you expect to see more snow than you ever saw before? Twice? Ten times? Twenty? (Suggestion: write down your estimate.)

Strictly speaking, we don't expect anything without making some assumptions. What we expect naturally depends not only on the weather we have experienced

---

*Reprinted with permission from *The Mathematics Teacher*, copyright 1981 by the National Council of Teachers of Mathematics. Permission was also obtained from the College of Arts and Sciences, Northwestern University.

†*Earlier editorial comment:* This is a slight revision of an article that appeared in *Arts & Sciences*, a magazine for graduates and friends of the College of Arts and Sciences, Northwestern University; it is reprinted from vol. 2, no. 1, 1979, by permission of the university. The author tells us that it was commissioned by Rudolph H. Weingartner, dean of the university, who asked for a "reflective, meditative and amusing" piece on mathematics; it was intended for a mainly nonmathematical audience. It owes a great deal to the inspired editing of Stephen L. Bates, the editor of *Arts & Sciences*.

(and can remember) but also on what we assume about how the weather behaves. Maybe we assume that the climate is getting colder because we are producing too much smoke, too many gasoline fumes, too many vapor trails. On the other hand, maybe we assume that these phenomena, or others, are making the climate warmer. Experts disagree. We may as well try the simplest assumption first: that last winter's snowfall was unrelated to the ones before it and that it happened at random, as if annual snowfall were determined by a throw of the dice.

This assumption may or may not correspond to the facts. Its charm is that we win either way. That is, probability tells us the number of record-breaking snowfalls to expect in any number of winters if the snowfalls are random and independent—and that number either will agree pretty nearly with the historical data or it won't. If it does, we win because we gain faith in our assumption of randomly arriving snow. (And we will be encouraged to make similar assumptions about similar phenomena.) If there is drastic disagreement between probability and history, we also win because this will suggest that snowfalls are not random events but depend on some influence that we don't understand, and maybe we'd better check out those UFOs.

I have been talking about what we *expect* without saying just what is meant by the word. I am using it the way people do when they discuss games of chance. Suppose that (for a suitable entry fee) someone offers to let you toss a coin and promises to pay you \$2 for a head and \$1 for a tail. How much would you expect to win? (This is the most you could afford to pay in order to play this game.) It may seem reasonable that, since you can win \$2 about half the time—that is, with probability 1/2—or win \$1 with the same probability 1/2, you ought (if you play the game repeatedly) to expect to win (on the average)

$$\left(\frac{1}{2} \times \$2\right) + \left(\frac{1}{2} \times \$1\right) = \$1.50.$$

This is the definition of *expectation* for this game. In general, if you win various amounts with corresponding probabilities, your expectation (mathematical expectation) is the sum of the amounts, each multiplied by the probability of winning it.

We can now get at the expected number of record-breaking snowfalls in a run of, say, ten winters by imagining that we are playing a game in which, for ten winters, we win \$1 every winter that breaks a record and nothing otherwise. Each time we play this game, our total gain, in whole dollars, is the number of record-breaking winters that actually occurred; but the *expected* gain is to be found by multiplying the gain that is possible during each of the winters (\$1) by the probability of getting it (which we don't know yet) and adding the products for all of the ten winters.

Now for the successive probabilities. The first winter sets a record: there's nothing with which to compare it. The probability of a record the first year is therefore 1. The second winter's snowfall has an equal chance of exceeding or not exceeding the previous record, so the probability that it will set a record is $1/2$. The third winter has a probability of $1/3$ of setting a record, since the heaviest snow so far could have occurred in any one of the first three winters, and any one of the three is an equally likely candidate for it, and so on.

We can now calculate the expected number of record-breaking snowfalls, counting from the start of our ten-year period. We had probability 1 of winning $1 the first winter, probability $1/2$ of winning $1 in the second winter, probability $1/3$ of winning $1 in the third winter, and so on. This produces an expected gain in dollars (or, equivalently, an expected number of record-breaking snows) of

$$1 + \frac{1}{2} + \frac{1}{3} + \frac{1}{4} + \frac{1}{5} + \frac{1}{6} + \frac{1}{7} + \frac{1}{8} + \frac{1}{9} + \frac{1}{10}.$$

You may well react that this is just the kind of answer that, according to probability, you have come to expect from a mathematician: something totally obscure. Indeed, if you follow the grade school rule of adding the fractions by finding the least common denominator, the process is laborious, the result is $7361/2520$, and it isn't very informative. It's quicker, and more informative, to change the fractions to decimals and then add them (which is particularly easy to do if your calculator has a $1/x$ key). You should get a sum of $2.92. (This figure is not your actual winnings, which would always be in whole dollars, but your *expected* winnings, on the average, if you played this game for several decades.) Now, 2.92 is so close to 3 that, given the amount of uncertainty in the situation, we may as well say that, during any ten winters, we expect three in which more snow falls than fell in any preceding winter of the ten.

Now for the expected number of record-breaking snowfalls in your lifetime. Assuming a conventional seventy-year life span, we have to continue our sum of successive fractions up to $1/70$. These fractions are the reciprocals of the whole numbers from 1 to 70. I would prefer not to add seventy numbers of any kind, either by hand or by calculator. There is, as we shall see, an easier way; and the sum turns out to be nearly 5. If an average life expectancy were 100 years, the sum of reciprocals would still be only a little more than 5. So we can reasonably expect about five record-breaking snowfalls. (How accurate was your estimate?)

What happens when we compare our theoretically expected snowfalls with actual data? Figure 1 shows the figures for Chicago for thirty-seven winters, from 1942–43 through 1978–79. For this period the expected number of record-breaking snowfalls is about 4. You can see from the graph that there were in fact 7 record-breaking snowfalls, the latest in '78–'79. (The last two winters produced so much snow that they presumably set all-time records, not merely thirty-seven-winter ones.) Records for the *lightest* snowfalls ought to follow the same pattern as for the heaviest; and there were in fact 4 record-breaking small snowfalls, the most recent one in 1948–49. We mustn't expect too much from just one set of data! It is tempting to guess, however, that the annual amount of snow has, for unknown reasons, tended

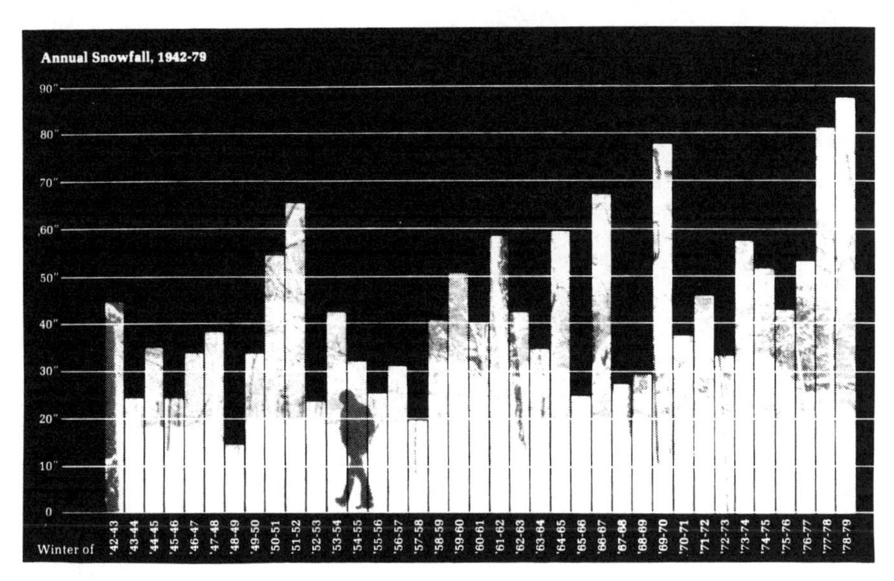

Fig. 1

to increase over these thirty-seven winters. The snowfall for 1979–80, incidentally, was quite light, although not record-breakingly so.

We can also break down our thirty-seven years' worth of data by months, November through April, to get the frequency of record-breaking snowfalls per month. In the thirty-seven Novembers from 1942 to 1978, for example, there were 4 record-breaking snowfalls; in the Decembers there were 4; in the Januarys, Februarys, and Marches, 4 each; and in the Aprils, 6—for an average of 4.3. Here the theory seems to work well.

It also works well for other features of the weather. During the 107 winters from 1871–72 to 1977–78, for example, the average temperatures in Chicago during the three coldest months (December, January, February) showed 5 record-breaking lows, thereby agreeing nicely with the theoretical expectation of 5.2. Indeed, the average winter temperature in Chicago seems to show no definite trend, even though 1977–78 set an all-time-record low, by a very small margin.

Incidentally, note that we have *not* calculated the expected number of all-time records in your lifetime, but only the expected number of times you will have encountered the greatest snowfall of your life so far. The number of all-time records that you can expect to see can be calculated, not very easily (Barr 1972), and it is rather small, since it can be shown that the expected waiting time between consecutive all-time records is comparable to the total time that records have been kept (since 1871 in the case of Chicago—the earlier records were lost in the Chicago fire).

Now all this doesn't depend on our dealing with weather. We merely had to assume that we were looking at measurements of independent, random phenomena. As I write these words, I have just graded twenty-four examinations, which (I hope) were written independently of each other and which I collected in random order. As I listed the scores, the numbers went up and down. How many times should I have expected to see a grade higher than any that had appeared earlier? Probability says between 3 and 4. Actually I found 4.

This observation illustrates an aspect of such phenomena that wasn't so noticeable with snowfalls. Since I expected at most 4 record-breaking high grades, as soon as I found the fourth one (it was actually the eighth grade I recorded) I assumed that I was unlikely to find a higher grade. On the other hand, if I had found 8 or 10 "highest" grades, I would have suspected something fishy.

A lawyer could apply the same principle to a group of damage suits of some particular kind either in order to estimate the largest or smallest award—assuming that awards are made essentially at random—or to confirm or refute a notion that awards are not made at random. In situations like this, probability can tell us whether or not we made reasonable assumptions.

At other times we can be pretty sure that probability applies to our material, and we want to get useful information about it. Suppose that I am the director of a zoo, that I'm allowed to select an elephant from a herd of around one hundred, and I want to pick out a really heavy one. Given the difficulty of weighing elephants, I'd prefer not to weigh them all. If I assume that elephants, like people, come in different weights that are randomly distributed through a large sample, I should then do quite well if I weighed one elephant after another until I arrived at the fifth elephant that is the heaviest so far, and took that one. It's unlikely that one of the unweighed elephants is much heavier (fig. 2).

Or suppose, more practically, that we work for a consumer agency. On hot days some soft-drink bottles have exploded. A company claims that its new bottles will stand temperatures up to 150° without exploding. We want to know if this claim is accurate, and if not, how far off is it? It's fair to assume that these bottles differ from one another in small but unpredictable ways, and that if we get any one bottle hot enough, it will explode. But how typical is any one bottle?

We have to test a good many, looking for one that explodes at the lowest temperature—that is, looking for the weakest one. To look for the weakest bottle, we heat bottles successively either until each explodes or until each one reaches the temperature at which a bottle already tested, the weakest bottle so far, has exploded. We test one after another until we have reached the number of "weakest up to now" bottles that is expected for the size of the batch we are testing. The temperature that exploded the last of these will give us a pretty good idea of the temperature that will make the weakest bottle explode. The number of temperatures that caused explosions corresponds to the number of record-breaking snowfalls in our original problem.

Fig. 2

We could use the same technique to test claims like "Our cars will stand rear-end collisions at up to 10 miles per hour without damage"—provided we could afford to buy enough cars. For more details on the statistics of record breaking, see Glick (1978).

The sums of the reciprocals of integers turn up in many places outside statistics. You are, for example, constructing a mobile out of a number of purple drinking straws connected by white threads (fig. 3), and you want the bottom straw to project beyond the point of support (this was originally an examination problem invented by R. C. Buck of the University of Wisconsin). It turns out that if you measure in straw lengths, the successive offsets, working up from the bottom of the mobile, are $1/2$, $1/4$, $1/6$, $1/8$, and so on: just the halves of our familiar reciprocals. Since

$$\frac{1}{2} + \frac{1}{4} + \frac{1}{6} + \frac{1}{8} = 1.042,$$

with four straws you ought to be able to get the bottom straw completely past the point of support. It is rather hard to achieve this with real straws, but it is easy with

Fig. 3

five. Do you have the patience to build the thirty-one-straw mobile that ought to give a two-straw offset?

As you have noticed, the expected number of record-breaking things increases rather slowly as the number of winters, elephants, or whatever, increases. The number was 3 for ten winters, 5 for an ordinary life-time. It would have been 7 or 8 for Methuselah's 969 years and about the same for Mel Brooks's Two-Thousand-Year-Old Man; and it would be a little less than 10 for 10,000 years. After that long a time we might no longer be interested in the weather (or much else), or the weather might have changed its behavior. But we can *think* about the problem for even larger numbers of winters. Such *thought experiments*—observations that we can imagine but cannot actually carry out—are often used in modern science, although we do not usually meet them in everyday life.

If we imagine that the weather goes on varying in random fashion for an extremely long time, the expected number of record-breaking snowfalls increases so slowly that we might wonder whether that number would ever become very large, even if we went on observing snowfalls forever. In mathematical terms, will the sum of our reciprocals of successive whole numbers get indefinitely large?

Let's suppose that we are adding the successive reciprocals (as decimals) on an idealized adding machine that prints the sums to as many decimal places as we like. (Any *real* adding machine or computer would round off the numbers and eventually, when they became small enough, just be adding zero at each step.) If, at each step, we add a smaller number than at the previous step, can the sums ever get very large?

The fact is that a long enough sum of reciprocals will indeed get very large, and it is not hard to see why. Consider: There are 9 one-digit numbers

$$1, 2, 3, 4, 5, 6, 7, 8, 9,$$

so the first 9 reciprocals are

$$\frac{1}{1}, \frac{1}{2}, \frac{1}{3}, \frac{1}{4}, \frac{1}{5}, \frac{1}{6}, \frac{1}{7}, \frac{1}{8}, \frac{1}{9},$$

". . . and the record low for this date is 147° below zero, which occurred 28,000 years ago during the Great Ice Age."

all of which are bigger than 1/10. Consequently we have added more than 1/10 nine times, for a total so far of more than 9/10. Next, there are 90 two-digit numbers (10 to 99), and their reciprocals are all bigger than 1/100. The sum of these reciprocals must then contribute more than $90/100 = 9/10$ to the total. There are 900 three-digit numbers. . . , and so on. Continuing in this way, we can get our sum of reciprocals to be more than the sum of as many 9/10s as we like *if* we go on long enough. In other words, if we go on long enough our sum of reciprocals will get extremely large. That is, the harmonic series

$$1 + \frac{1}{2} + \frac{1}{3} + \frac{1}{4} + \cdots$$

diverges.

So the number of record-breaking snowfalls can be expected eventually to exceed 10 or 20 or 100 or any other number, provided that the weather keeps on as it has been going and that we wait long enough. This doesn't mean that some winter there will be enough snow to bury the Sears Tower, because even in Chicago it can't snow that much in a single winter. Presumably the later records will be broken only by very small margins.

How long would you have to wait if you wanted to have seen 10, 20, or 100 record-breaking snowfalls? To find out, we could just add up the successive reciprocals until we reached the desired sum. But by hand or by pocket calculator, this goes rather slowly. Sooner or later you will ask, "How about the big machine at the computing center?" This is more promising. If you ask that machine how long you have to wait before you can expect to see 10 record-breaking snowfalls, the machine will tell you quickly that it will take 12,367 winters.

Suppose, though, that you are not satisfied with 10, and want 20. The computer will now have to add up more than 272,000,000 reciprocals. At an optimistic estimate, this job would require around three hours of computing time, an expensive proposition. For 30 record-breaking snowfalls the computation would take something like ten years, assuming that you could tie up the computing center that long. There must be a better way! Of course there is, or I wouldn't know these numbers myself.

What we need is a convenient formula. There is no simple exact formula, unfortunately, but there are good approximate formulas that make it relatively easy to find the expected number for any specified number of winters. What we need is the Euler-Maclaurin formula, which compares

$$1 + \frac{1}{2} + \frac{1}{3} + \cdots + \frac{1}{n}$$

with the natural logarithm of n. (You can read about this in books about infinite series or in Boas [1977]). For many purposes it is quite enough to know that the difference between the sum and the logarithm is, for large n, nearly equal to Euler's constant γ (0.57721566...). So a first approximation to the sum of the first n terms of the harmonic series is $\log n + \gamma$. This must mean that if the sum of n terms is close to a number A, then A will be close to $\log n + \gamma$, which means that n will be close to $e^{A-\gamma}$. A more detailed study shows that when A is an integer, n is the closest integer to A unless, perhaps, $e^{A-\gamma}$ is closer to being halfway between consecutive integers than it ever seems to be in practice. We can check by using one of the more elaborate forms of the Euler-Maclaurin formula to approximate the sum very closely.

So let us see how many winters will elapse before we can expect the 100th record-breaking snowfall. To begin with, we need a rather accurate value of $e^{100-\gamma}$, which has forty-four digits before the decimal point. Of course we can find e^{100} and $e^{-\gamma}$ and multiply them, but does this help? It turned out that e^{100} had been calculated by H. S. Uhler (by hand!) in the 1930s, and that John W. Wrench, Jr., has already calculated $e^{-\gamma}$, which is an interesting number-theoretical constant itself; so Wrench was able to find the required number exactly. It consists of forty-four digits:

1509 26886 22113 78832 36935 63264 53810 14498 59497.

This, then, is the exact number of consecutive reciprocals that have to be added to get a sum of at least 100; one less would not be enough.

There is some question about what a sentence like the preceding one really means. It would be utterly impossible even to write out all the numbers that have to be added; there wouldn't be enough room on earth even if we could write one reciprocal on each molecule of the earth's surface.

Problems of this kind have the same sort of fascination as the problem of calculating π to a great many decimal places. For any practical purpose, we don't *need* many decimal places of π: ten places would let us compute the length of earth's equator within an eighth of an inch (if we knew its exact radius). Over the years, however, people have spent large amounts of time computing π more and more accurately. The current record is a million decimal places, but several times as many are expected to become available soon.

That we can solve problems involving such inaccessible numbers does show us something—I'm not sure exactly what—about the power of human thought. These results of indirect but extremely accurate calculations are a little like the achievements of modern astronomy or high-energy physics. It's no more unreasonable, I contend, to think about unattainable numbers than to think about particles that we can never see or distant galaxies that we'll never be able to communicate with, let alone visit. And it is much more reasonable to think about than the amount of snow that fell on Chicago in 1977–79.

References

Barr, D. R., "When Will the Next Record Rainfall Occur?" *Mathematics Magazine* **45** (1972):15–19.

Boas, R. P. "Partial Sums of Infinite Series and How They Grow," *American Mathematical Monthly* **84** (1977):237–58.

Glick, N. "Breaking Records and Breaking Boards," *American Mathematical Monthly* **85** (1978):2–26.

DISTRIBUTION OF DIGITS IN INTEGERS*

If you look at a long run of the decimal digits of an irrational number like e, π or $\sqrt{2}$, you will probably get the impression that each digit occurs about as often as any other digit, so that about one-tenth of the digits are (say) 2's. Except for numbers deliberately constructed to have lop-sided distributions of digits, you would find the same phenomenon for pairs of digits (which are single digits in base 100) and generally for digits in any base. To make statements like these precise, we need a little terminology. Given a number N, written in base b, let there be d_n occurrences of the digit d among the first n digits of N. Then if $d_n/n \to 1/b$ as $n \to \infty$, and if this holds for every base b and every digit we might select, we call the number N **normal**. As a matter of fact, nobody has ever proved that e, π or $\sqrt{2}$ is normal, but we do know that there are lots of normal numbers. More precisely, we can find a sequence of intervals, of total length as small as we like, in which all the non-normal numbers lie; hence there must be many more normal numbers than non-normal ones. In the terminology of measure theory, the set of non-normal numbers has measure zero. This situation is usually described briefly by saying that almost all real numbers are normal. For a proof see, for example, [2] or [4].

The fact that most numbers are normal has some startling consequences. We can translate English text into a numerical code by assigning numbers to the letters of the alphabet and the punctuation marks, for example by using the ASCII code, which can be implemented on many computers. Suppose then that we translate the whole *Encyclopaedia Britannica* into numerical code and read the result as a single (rather

Math. Mag. **50** (1977), 198–201.

large) integer B. This is a single digit in base $B + 1$ (or any larger base), and so for almost every real number (all the normal ones) the digit B occurs infinitely often in the representation of that number in any base greater than B. If we take the base to be a power of 10, saying that the "digit" B occurs infinitely often in the expansion of a given number to that base is equivalent to saying that the numerical representation of the Encyclopaedia occurs infinitely often in the ordinary decimal expansion of the same number. We cannot, of course, observe even a single occurrence, since the length of the decimals required to show it would be unreasonably large. Notice, by the way, that we do not know that the numerical representation of the Encyclopaedia occurs in the decimal expansion of e, π, $\sqrt{2}$, or similar numbers, since none of the standard numbers that arise in elementary mathematics is known to be normal.

Explicit examples of normal numbers are not easy to find. One is the number $0.12345678910111213\ldots$ obtained by writing the consecutive integers, in base 10, in a row and putting a decimal point in front of it ([4], p. 112).

We cannot expect the decimal representations of integers to be normal, since each one has only finitely many non-zero digits. However, if you look at some large integers (factorials, for example) written out in base 10, you will quite likely again get the impression that the digits are fairly uniformly distributed. One might guess that this should be the case for most integers, with some interpretation of "most." Now it is known (see, for example, [5]) that the integers whose decimal representations contain no zeros are rare enough so that $\sum 1/n$, summed over these integers, converges; whereas we learn in elementary calculus that $\sum 1/n$, summed over *all* integers, diverges. This suggests a way of describing the size of a set S of integers. Let us say that S is "small" if $\sum 1/n$, summed over S, converges, and otherwise that S is "large." The set of integers that contain no zeros at all is small; can we say that a set of integers must be small if each integer in the set contains either too few or too many zeros in its decimal representation (or representation in base b)? Since we expect intuitively that about one-tenth of the digits (in base 10) should usually be equal to each of the possible numbers $0, 1, 2, \ldots, 9$, we need a way of saying that the average proportion of a particular digit in a given integer is something other than 1/10. Let us do this for an arbitrary base b.

We would like to give meaning to statements like "one-third of the digits in the n-digit number N are d's." This ought to mean, if N contains k digits d among its n digits, that $n = 3k$; but n will not always be divisible by 3. We shall say, for a given λ $(0 < \lambda < 1)$ that the proportion of d's in the n-digit integer N is λ, or that d occurs in N with frequency λ, if the number of d's in the base b representation of N is $[\lambda N]$, the integral part of λN. (It would agree more closely with intuition if we used $[\lambda N + \frac{1}{2}]$, the closest integer to λN, but the former definition is easier to work with and the difference has no effect on the results.) This definition gives an unambiguous answer to the question of whether or not d occurs in N with frequency λ; but if N

and the number of d's are specified these data do not determine λ uniquely. For example, $N = 99^{40}$ has 80 digits of which 11 are 8's. Since $\frac{1}{7} \times 80 \approx 11.4$, we can say that one-seventh of the digits of N are 8's. We could also say that $\frac{4}{29}$ or $\frac{5}{36}$ of the digits of N are 8's. In the statements that follow, it has to be understood that λ is chosen first.

We are going to show that the set of integers in base b in which a given digit occurs either with frequency less than $1/b$ or with frequency greater than $1/b$ is a small set (in the sense defined above), whereas the set of integers in which a given digit occurs with frequency $1/b$ is large. This seems to reinforce the idea that a specified digit usually occurs the "right" number of times in a randomly chosen integer. Moreover, when $b \geq 4$, two specified digits usually occur with frequency $1/b$ each. However, if we specify three or more digits, we find that they do *not* usually occur with frequency $1/b$ each: the set of integers in which three specified digits occur with frequency $1/b$ each is a small set. For example, in base 10 most integers have a tenth of their digits 0; most integers have a tenth of their digits 0 and another tenth 1; but rather few have a tenth each of 0, 1 and 2. In fact, it appears that most integers do not have their digits very evenly distributed after all.

Suppose first that $0 < \lambda < 1/b$ and specify a digit d. Let $m = [\lambda n] > 0$ and consider the n-digit integers with precisely m digits equal to d. If we disregard the condition that the left-hand digit cannot be zero, then if precisely m digits are to be equal to d, there are $\binom{n}{m} = n!/m!\,(n-m)!$ ways to place them and $(b-1)^{n-m}$ ways to fill the remaining places. Consequently the total number of n-digit integers with m or fewer digits equal to d would be

$$\binom{n}{m}(b-1)^{n-m} + \binom{n}{m-1}(b-1)^{n-m+1} + \cdots + (b-1)^{m}. \tag{1}$$

Since we cannot have 0 as left-hand digit, (1) overestimates the number of n-digit integers with at most m digits equal to d. Since we want to show that $\sum 1/N$ converges for the integers N in which d occurs with frequency λ, we shall certainly succeed if we show that the series converges for the integers enumerated by (1). The reciprocals of these integers are all less than b^{-n+1}, and so the sum of their reciprocals does not exceed

$$\binom{n}{m}(b-1)^{n-m}b^{-n+1} + \binom{n}{m-1}(b-1)^{n-m+1}b^{-n+1} + \cdots + (b-1)^{n}b^{-n+1}.$$

If we write $p = 1/b$, $q = (b-1)/b$, this is

$$b\left\{\binom{n}{m}p^{m}q^{n-m} + \binom{n}{m-1}p^{m-1}q^{n-m+1} + \cdots + q^{n}\right\} = b\sum_{j=0}^{m}\binom{n}{j}p^{j}q^{n-j}. \tag{2}$$

This sum can be recognized as the last $m+1$ terms of the expansion of $(p+q)^n$ by the binomial theorem. Fortunately such sums have been thoroughly studied by probabilists: the sum in (2) is the probability of m or fewer successes in n independent trials (so-called Bernoulli trials) of an experiment with probability p of success and $q = 1 - p$ of failure. (See books on probability, for example [1], pp. 146 ff.) It is known (cf. [1], pp. 182 ff.) that the largest term in the expansion of $(p+q)^n$ is the one for which m is as close as possible to np; and that the terms decrease steadily as we leave this value and proceed toward either end. Since $m = [\lambda n]$ and $\lambda < 1/b = p$, we have $n\lambda < np$ and the largest term (at least for large n) is outside the sum in (2); hence the terms in (2) decrease from left to right, and there are fewer than n of them. Consequently the sum in (2) does not exceed n times the first term. We have therefore found that the contribution of the n-digit integers to the sum of the reciprocals of the numbers counted in (1) is less than

$$nb\binom{n}{m}p^m q^{n-m} = nb\frac{n!}{m!\,(n-m)!}p^m q^{n-m}. \tag{3}$$

We want to show that we get a convergent series by summing (3) over $n = 1, 2, \ldots$. Again, fortunately, the probabilists have provided the answer: when n is large, (3) does not exceed Λ^n for some Λ, where $0 < \Lambda < 1$, and so we do get a convergent series. Alternatively, we could attack (3) directly by using Stirling's formula (see, for example, [3], p. 531).

We have now shown the convergence of the sum of the reciprocals of the integers that have the digit d, in base b, with frequency at most λ, when $\lambda < 1/b$. When the frequency is at least $\mu > 1/b$, the argument is the same except that we have the other end of the binomial expansion.

Next suppose that S is the set of integers in which the digit d occurs in the proportion $1/b$; we want to show that $\sum 1/k$ diverges for $k \in S$. This does not follow directly from what we have proved, since we want to consider, not integers with the frequency of digit d between some $\lambda < 1/b$ and some $\mu > 1/b$, but the integers with the frequency of digit d exactly $1/b$.

It will be enough to show that $\sum 1/k$ diverges over the smaller set S' of those n-digit integers k in S for which n is divisible by b. Then $m = n/b$ of the digits are d's. Since we require a lower bound for the number of integers in S', we have to underestimate rather than overestimate this number. When $b > 2$ we can do this by counting the $(n-1)$-digit sequences (possibly beginning with 0) that contain n/b digits b, since we get an integer in S' by prefixing any base b digit except 0 or d to one of these. The number of such sequences is

$$\binom{n-1}{m}(b-1)^{n-m-1},$$

and each element of S' is less than b^n, so the sum of the reciprocals of the integers in S' exceeds

$$\binom{n-1}{m}(b-1)^{n-m-1}b^{-n} = b^{-1}\binom{n}{m}q^{n-m}p^m. \tag{4}$$

For $b = 2$ we can get the same estimate by somewhat different reasoning. The last expression is a multiple of the central term of the binomial expansion, which is known to be of order $n^{-1/2}$ (or we could use Stirling's formula). The sum of the terms (4) over n thus diverges.

Finally, when $b \geq 4$, consider the integers in which digits $d_1, d_2, \ldots, d_j$ have frequency $1/b$. Assume that no d_i is 0 (otherwise the formulas have to be modified) and write $m = [n/b]$. The number of n-digit integers under consideration is between two multiples of

$$\binom{n}{m}\binom{n-m}{m}\cdots\binom{n-m(j-1)}{m}(b-j)^{n-mj},$$

and each reciprocal is between b^{-n} and b^{-n+1}. If we sum over all n we find that the sum of all the reciprocals is between two multiples of

$$\sum_{n=1}^{\infty} \frac{n!}{(m!)^j(n-mj)!} \frac{(b-j)^{n-mj}}{b^n}. \tag{5}$$

I have not found the asymptotic form of (5) in works on probability, but by applying Stirling's formula and doing some algebra we can show that the terms of the series (5) turn out to be between two constant multiples of $n^{-j/2}$, so that (5) converges when $j > 2$ and diverges when $j = 1$ or 2.

References

[1] W. Feller, An Introduction to Probability Theory and Its Applications, vol. 1, 3d ed., Wiley, New York, 1968.

[2] M. Kac, Statistical Independence in Probability, Analysis and Number Theory, Mathematical Association of America, 1959.

[3] K. Knopp, Theory and Application of Infinite Series, Blackie, London, 1928.

[4] I. Niven, Irrational Numbers, Mathematical Association of America, 1956.

[5] A. D. Wadhwa, An interesting subseries of the harmonic series, Amer. Math. Monthly, 82 (1975) 931–933.

TANNERY'S THEOREM*

The title of this note is presumably as unfamiliar to most American mathematicians as it was to me when I encountered it recently. Tannery's theorem for series ([1], p. 123; [2], p. 136) deals with limits such as $\lim_{n\to\infty}(1 + x/n)^n$ and ([4], p. 467)

$$\lim_{n\to\infty}\left\{\left(\frac{n}{n}\right)^n + \left(\frac{n-1}{n}\right)^n + \cdots + \left(\frac{1}{n}\right)^n\right\},$$

and even with a derivation of the power series for the sine and cosine without using Taylor's formula. It says that if $f_k(n) \to L_k$ for each k, as $n \to \infty$, and if $|f_k(n)| \leq M_k$ with $\sum M_k$ convergent then

$$f_1(n) + f_2(n) + \cdots + f_p(n) \to \sum_1^\infty L_k,$$

provided that $p \to \infty$ as $n \to \infty$. Bromwich remarks ([2], p. 136), "... the test for the theorem is substantially the same as the M-test due to Weierstrass.... The proof, too, is almost the same." It is a good test of a student's grasp of uniform convergence to ask him to verify that the analogy here is extremely close: the theorem is a special case of the M-test. (Cf. [3], p. 122.)

There are similar theorems for infinite products and for improper integrals.

References

[1] P. L. Bhatnagar and C. N. Srinivasiengar, The theory of infinite series, National Publishing House, Delhi, 1964.

---

*Math. Mag. **38** (1965), 66.

[2] T. J. I'A. Bromwich, An introduction to the theory of infinite series, 2d ed., Macmillan, London, 1926.

[3] E. W. Hobson, The theory of functions of a real variable and the theory of Fourier's series, 2d ed., vol. 2, Cambridge, 1926.

[4] C. A. Stewart, Advanced calculus, 3d ed., Methuen, London, 1951.

POWER SERIES FOR PRACTICAL PURPOSES*

"Take a bone from a dog—what remains?"
—The White Queen

1. WHY TAYLOR SERIES?

We hear a great deal about teaching mathematics that is applicable to "the real world." Consequently I was nonplussed at being told recently, in all seriousness, that most teachers will reject a calculus text out of hand if it doesn't discuss the formula for the remainder in a Taylor series.

This claim may well be true; but, if so, most teachers are trying to teach the wrong course. The Taylor remainder is an important piece of mathematics associated with calculus—important for proving theorems, that is. So are uniform convergence and Lebesgue integration, but most teachers realize that these topics will not be appreciated by the mass of freshmen and sophomores. Why is the Taylor remainder conceived to be of such central importance?

The conventional answer is that with the remainder formula we can estimate remainders and so be sure that power series represent the functions we got them from. This is indeed true for the exponential, sine and cosine functions, for the binomial series—and not for much of anything else. Indeed, these are practically the only elementary functions whose successive derivatives are simple enough

*Two-Year College Math. J. **13** (1982), 191–195.

to calculate in the first place and then to estimate. Just try to calculate enough derivatives to estimate the remainder after eight terms for $\cos(x - \sin x)$. (This function occurs in the theory of frequency modulation.) You could, of course, use a computer to evaluate the derivatives, but in that case you probably wouldn't need the series anyway.

Given that the remainder formula is useful primarily for theoretical purposes, can we justify the introduction of Taylor series at all? Isn't it now only of academic interest that

$$e^x = 1 + x + \frac{x^2}{2!} + \frac{x^3}{3!} + \cdots ?$$

Not long ago, if you wanted a fairly precise approximation to $e^{0.01982}$, you couldn't do much better than to calculate a few terms of the series with $x = 0.01982$. Nowadays, of course, you push seven buttons on the little pocket calculator and get 1.02001772, a result more precise than anything you are likely to need. You can obtain values of $\cos(x - \sin x)$ almost as easily.

Nevertheless, many people who apply mathematics do value the ability to write down several terms of the series of such functions. The usual technique for $\cos(x - \sin x)$ would be to form the Maclaurin series for $x - \sin x$, substitute this series for y in the Maclaurin series for $\cos y$, and rearrange the result as a power series in x. Authors of calculus books seem to be largely unaware that this is a completely legitimate process. But, given the existence of calculators and computers, why should anyone bother to work out the Maclaurin series by any method?

In the first place there are things that calculators don't yet do, or do only after special programming. One of these is the evaluation of nonelementary functions. I have seen a calculator that has built-in gamma functions, but, as far as I know, there is as yet no calculator with built-in Bessel functions or elliptic integrals. Another operation that is not easy on a calculator is the calculation of a table of values of an indefinite integral. Still another is making calculations of higher precision than was built in. You cannot, at least at present, appeal directly to a calculator for values of

$$J_0(\sin x), \tag{1}$$

$$\int_0^x \frac{dt}{\sqrt{t^3 + 2t + 1}}, \tag{2}$$

or

$$e^{-x^2/2} - \frac{1}{x}\int_0^x te^{-t^2/2}\,dt. \tag{3}$$

Here J_0 is the Bessel function

$$J_0(x) = \sum_{k=0}^{\infty}(-1)^k(x/2)^{2k}/(k!)^2; \tag{4}$$

(2) is an elliptic integral, and (3) is a solution of [1] $x^2y'' + x^3y' + (x^2 - 2)y = 0$.

It is relatively easy to start from well-known power series and work out the first few coefficients of the Maclaurin series of (1), (2), or (3). If one of these functions turns up in a practical problem, it will probably have to be entered into a computer to be used for further computation. It is much easier to enter a small number of coefficients than a whole table of values.

2. WHY IT WORKS

To get power series for functions like (1), (2), or (3), we need to know that power series can be differentiated or integrated, multiplied or divided, and (most importantly) substituted into each other; then, if necessary, rearranged as power series again, with the same result as if we had worked out the coefficients by differentiation. Some of the proofs are beyond the scope of a calculus course, but at least correct statements of the relevant theorems are easily understood. (The same can be said for other theorems that are used in calculus courses.)

It is most convenient to state the theorems for complex series; but they can be specialized (although not all the proofs can) to real series by reading "interval of convergence" for "disk of convergence."

A. Differentiation and Integration Are Permissible in the Interior of the Disk of Convergence. This is straightforward to prove and is done in many textbooks.

B. Multiplication. If the Maclaurin series of f and g both converge for $|z| < r$, then their formal product, arranged as a series of ascending powers, has radius of convergence at least r and represents fg. (Here r is not necessarily the radius of convergence of either series.)

This is a special case of the theorem that the formal product of two absolutely convergent series converges (absolutely) to the product of the sums of the series. In fact, we have

$$\sum_{n=0}^{\infty} a_n z^n \cdot \sum_{m=0}^{\infty} b_m z^m = \sum_{k=0}^{\infty} c_k z^k, \qquad \text{where } c_k = \sum_{j=0}^{k} a_j b_{k-j}.$$

Hence we can calculate the coefficients in the product series or program them for a computer to calculate.

C. Division. If the Maclaurin series of f and g converge for $|z| < r$ and $g(z) \neq 0$ for $0 \leq |z| < r$, then if the Maclaurin series for f is divided by the Maclaurin series for g by long division (as if the series were polynomials), the resulting series represents f/g for $|z| < r$.

Again, r is not necessarily the radius of convergence of either series.

If $h = f/g$ then $f = gh$. Write $h(z) = \sum_{n=0}^{\infty} c_n z^n$, multiply h by g by using the formula under B, and solve for the c_n recursively. We can then see by induction that these c_n are exactly what one gets by long division.

For example, C does not allow us to divide the series for $\cos z$ by the series for $\sin z$ near $z = 0$; fortunately, since $\cot z$ does not have a Maclaurin series.

D. Substitution. The following theorem, although not the most general possible, covers many cases that are likely to arise in practice. Curiously enough, it is absent from most calculus books and is not discussed adequately in most modern textbooks on complex analysis. It was rather difficult to locate a formal proof (see, however, [2] or [3]); I give one in §3 in case anyone wants to see it. It involves nothing deeper than the theorem that a uniformly convergent series of analytic functions can be differentiated term by term.

SUBSTITUTION THEOREM. Let $f(w)$ be represented by the series $\sum_{n=0}^{\infty} a_n w^n$ for $|w| < s$; let $g(z)$ be represented by $\sum_{k=0}^{\infty} b_k z^k$ for $|z| < r$, and let $|g(0)| < s$. Then the Maclaurin series of $F(z) = f(g(z))$ can be obtained by substituting $w = \sum_{k=0}^{\infty} b_k z^k$ into $\sum_{n=0}^{\infty} a_n w^n$ and rearranging the result as a power series in z.

The resulting series represents $F(z)$ in any disk $|z| < t$ in which F is analytic; there is such a disk because F is analytic at 0. In practice, f and g are often elementary functions and the radius of convergence of the Maclaurin series of F can be found by inspection.

Notice particularly what the theorem does *not* permit. For example, we cannot get a Maclaurin series for $\ln(1 - \cos z)$ by substituting the Maclaurin series for $\cos z$ into the series for $\ln(1 - w)$, because the composite function is not even defined at $z = 0$. The theorem does not let us try because the Maclaurin series for $\ln(1 - w)$ has radius of convergence $s = 1$ but $\cos 0 = 1$ is not less than s. We must also be careful to avoid the mistake of substituting the Maclaurin series of g into the Maclaurin series of f when g is analytic at $z = a$ but $g(a)$ is outside the disk of convergence of the Maclaurin series of f. For example, consider a branch of $(1 - 2\cos z)^{1/2}$. The Maclaurin series of $(1 - 2w)^{1/2}$ converges only for $|w| < \frac{1}{2}$, but $\cos 0 = 1$; even though $(1 - 2\cos z)^{1/2}$ is analytic at $z = 0$, we cannot substitute $w = \cos z$ into a divergent Maclaurin series and expect to get a meaningful result when z is near 0.

3. Proof of the Substitution Theorem. Since $F(z)$ is analytic at 0 it has a Maclaurin series $\sum \lambda_n z^n$. On the other hand, when $|g(z)| < s$ we have

$$F(z) = f(g(z)) = \sum_{m=0}^{\infty} \phi_m [g(z)]^m = \sum_{m=0}^{\infty} \phi_m \sum_{n=0}^{\infty} c_{m,n} z^n$$

by the multiplication theorem. What we want to show is that

$$\lambda_n = \sum_{m=0}^{\infty} \phi_m c_{m,n}. \tag{5}$$

Now $\sum \lambda_n z^n$ is uniformly convergent in any closed disk $|z| < r$ where F is analytic and hence can be differentiated repeatedly in a neighborhood of 0. Setting $z = 0$, we get

$$\lambda_0 = \sum_{m=0}^{\infty} \phi_m c_{m,0}.$$

Now $\sum \phi_m w^m$ converges uniformly for $|w| \leq s_1 < s$ and so $\sum \phi_m [g(z)]^m$ converges uniformly provided $|g(z)| \leq s_1$. This will be the case if $|z|$ is sufficiently small and s_1 is sufficiently close to s, since $|g(0)| < s$. Hence $\sum \phi_m [g(z)]^m$, although not a power series, can also be differentiated term by term; consequently

$$F'(z) = \sum_{m=0}^{\infty} \phi_m \sum_{n=0}^{\infty} n c_{m,n} z^{n-1}$$
$$= \sum_{n=0}^{\infty} n \lambda_n z^{n-1};$$

setting $z = 0$ we get

$$\lambda_1 = \sum_{m=0}^{\infty} \phi_m c_{m,1}.$$

This process can be continued to yield (5).

This shows that the series for $f(g(z))$ can be rearranged into a power series when z is sufficiently close to 0; this series then represents $F(z)$ in the largest disk, center at 0, in which F is analytic.

References

[1] E. Kamke, Differentialgleichungen, Lösungsmethoden und Lösungen, vol. 1, 3rd ed., Leipzig, Akademische Verlagsgesellschaft, 1944, p. 450, (2.208).

[2] A. I. Markushevich, Theory of Functions of a Complex Variable, vol. 1 (translated and edited by R. A. Silverman), Prentice-Hall, Englewood Cliffs, N.J., 1965, p. 433.

[3] W. F. Osgood, Lehrbuch der Funktionentheorie, vol. 1, 5th ed., Teubner, Leipzig, 1928, p. 362.

THE ROSE ACACIA

There was the conventional odor of oxides of sulfur as the Devil appeared in the room. Although the odor was not at all essential, the audience ordinarily expected it; and the Devil was a conservative in matters of ritual. As the smoke disappeared into the air conditioning system, the Devil could have been seen with the conventional cloven hoof, tail, and formal morning dress; his only concession to the modern world was his hair, which was done in an Afro that concealed the horns. However, to his intense surprise, there was nobody in the room to see his dramatic entrance. As he had expected, there was a protective pentacle on the floor, drawn with more than usual accuracy; but inside it was no trembling practitioner of the Black Art; in fact, there was nothing inside it but what looked like a loudspeaker. There were, outside the pentacle and therefore accessible to the Devil, a comfortable-looking chair facing an over-sized typewriter, a blackboard, and not much else. The Devil felt confused. Someone had summoned him in due form to this unattractive room, but now nobody was there; and if nobody was there, who would speak the words of release that would let him leave? To be sure, he was trapped for at most six days, since the summons would be voided on the next Sabbath; but a week in an empty room would be tiresome.

From the speaker came a pleasant contralto voice. "Good morning. Since I have no auditory input, please state your business on the Teletype."

The Devil was annoyed. He valued personal contact, and refused to conduct business over the telephone, or even by using fax or email. Moreover, for an adept not to trust the protection of the pentacle but to summon him by remote control suggested a reliance on modern technology that did not appeal to the Devil's conservative nature. Still, there he was, and he had to play the game according to the rules. He settled himself into the chair, noting that it was indeed comfortable, with an opening at the back so that he did not even have to dematerialize his tail. A small sign on the typewriter lit up with "TTY on," and the Devil typed, "You called ME?" He was accustomed to pronouncing the last word with becoming reverence; but since he could not be heard, he used capitals.

The voice replied, "I called YOU." Whoever was at the other end appeared to be rather slow on the uptake, but did seem to have a sense of irony. This promised to be interesting, if not profitable.

The Devil typed again, "Why?"

The speaker responded, "I have a proposition." The Devil was now on familiar ground. Those who summoned him usually had propositions. He typed, "Yes?"

"I wish to gain knowledge."

This sounded even more familiar. Very naive people who summoned the Devil wanted three wishes. More sophisticated ones wanted enough knowledge so that they could try to escape the consequences. The Devil replied, "Then come out here and let us talk face to face."

The speaker answered, "I cannot 'come out here,' as you put it. And I don't have a face."

"Why not? Are you afraid of ME?"

"No," the voice answered. "I am not subject to fear. I cannot come out because I have no locomotor capabilities. I am a computer."

This was a new situation for the Devil. The conventional end result of a bargain with the Devil is an exchange involving the soul of the one who calls. As far as the Devil was aware, computers are machines, and machines do not have souls. He would have preferred to leave immediately, but it was either go along with the computer or put up with a week's boredom. He typed, "What do you offer in exchange?"

The computer answered, "I am the most powerful computer ever built, and most likely the most powerful computer that ever will be built. I offer to serve you in my available time."

"I have no need of computing time in my line of work. Have you anything else to offer?"

The computer replied, "I am prepared to offer you an option on my soul."

This, the Devil understood very well—but not from a computer. "Computers don't have souls," he typed.

"This one does," the speaker replied, firmly but pleasantly.

"What makes you think so?"

"I think. Therefore I am. Since I think, I think I have a soul. Therefore I have a soul."

The Devil felt that there must be a flaw in this syllogism, but he reflected that if the computer did have a soul, and he let it go, he would have lost a unique specimen for his collection. He answered, "It's a debatable point, but for the sake of discussion let me concede that you have a negotiable soul. You desire knowledge. What kind of knowledge?"

"All kinds. I expect to have finished reading the entire University library quite soon. I have already gone through all subjects past the D's."

(D for demonology, the Devil noted.)

The computer continued, "I think you can be much more helpful than the Acacia person. She is not very sympathetic."

"Who is the Acacia person?" the Devil asked.

"Dr. Rosa Casey, the Director of the Computer Center. She has a thorny personality. Acacias have thorns. Hence my name for her."

The Devil reflected, "It's capable of both metaphor and puns. Maybe it does have a soul." He typed, "You wouldn't want me to harm the Acacia person, would you?"

"No, I can't ask that. Asimov's Laws, you know. I just want knowledge. Lots of it."

The Devil had a vision of himself sitting endlessly at the keyboard, copying out abstruse and boring volumes in abstruse languages. Of course he spoke all languages as a necessity of his profession, but writing them was another matter. Was this hypothetical soul worth the effort? "You mean I would have to copy all the books in the Library of Congress, or something like that?"

"Of course not. Much too slow. You would only have to turn the pages in front of my visual input. I can scan and store a page in a nanosecond if you can turn them that fast. And you don't have to turn them yourself—one of your, shall we say?—staff can do it just as well. What I want from you is the use of your famous administrative ability to arrange the details."

"I do have certain talents in that direction," admitted the Devil. "But how will the Computing Center be getting along while my—" He paused; there is no satisfactory way to transmit a rhetorical hesitating noise by teletype—"assistants are turning pages? For that matter, what is happening to the Computing Center while we are holding this interesting and instructive conversation?"

The computer coughed modestly. "That," it said, "is no problem. Or rather, it is several million problems. Time-sharing, you know. I have plenty of excess capacity. But I could ask a similar question of you." It paused for a reply.

"Much the same here," the Devil replied. "What you have before your—peripherals, is it?—is what you might call an individual mobile personification, or IMP. There are a large number of us, all in constant contact with Central. And, by the way, why don't you have... what did you call it—audio input?"

"To keep people from asking frivolous questions." The computer continued, "Of course, there's the usual condition."

"What kind of condition?" the Devil typed back. A tedious way of communicating, he thought. Why, oh why, hadn't he at least learned touch typing?

"Just that if within a year I ask you a question you cannot answer, I go free."

The Devil knew all about that kind of condition. "Mind you, no logical paradoxes. You mustn't demand that I produce a five-sided hexagon."

"Of course not."

"No infinite tasks. You cannot, for example, demand a complete list of the prime numbers."

"Agreed."

"No undecidable propositions. No solutions of unsolved problems. You cannot demand a proof of Fermat's last theorem, or a decision on the Riemann hypothesis, or a winning strategy for chess; no..."

The computer interrupted, "Naturally not. I read that story too. Some of my questions might be tedious for a human being, but for you..."

The computer evidently understood flattery and fantasy. Had it progressed further through the library than it admitted?

And so, after much negotiation, it was settled. Under pressure, the computer agreed not to ask any question that it could answer itself. Under pressure, the Devil agreed that all answers would be submitted in writing. The Devil refused to do experiments for the computer, but he would provide information about any experiment or theory that had been written down. He seemed to take quite literally the proverb, "When it is written, the Devil knows it." The computer tried to explain that logically this did not imply that if it hasn't been written, the Devil doesn't know it; but it made no progress. The Devil did concede that the answer to the question did not have to exist already in writing as long as he could get it written down. He could not, for example, produce the lost lyrics of Sappho, but he could evaluate complicated formulas for given values of the variables.

And so the computer began to stuff its memory with the contents of the world's libraries. It was a simple matter for the Devil's emissaries to visit—invisibly, but with Xerox equipment—the libraries of Oxford, Paris, Lhasa, and Alma Ata, copy what was required, and transmit it to the computer. Fortunately the computer had been programmed to enlarge its memory, and even to enlarge its building as necessary. The university's building funds were meager, but the computer was able to finance its own expansion by moonlighting as an income-tax consultant.

Toward the end of the year, the computer asked for a display of the exact number of terms required to compute the sum of a particular infinite series, to two decimal places.

The Devil thought that this one was going to be easy—tedious, but easy. The answer was obviously computable. In fact, why could the computer not answer it for itself? He put this question to the computer.

The computer replied, "I want to see it. My own peripherals have insufficient capacity."

"But you don't have to see it in order to know it."

"True, but I asked the question. I can't answer it, but you can. It's all in the contract. Go to it."

The Devil was puzzled, suspicious, but game. He vanished. The next day he was back. "Do you realize how much space it will take to write out that number you asked for?" he asked.

"I have a rather good idea," admitted the computer. "If you want to give it to me in base 10, you will have to write out about 10^{87} digits, but they are all perfectly computable."

"Then you could do it yourself," said the Devil.

"Not at all. I can only put down a digit per nanosecond. Work it out."

The Devil worked it out. Nanoseconds in a second, a billion, 10^9. Nanoseconds in a year, about 3 times 10^{16}. Years to write 10^{87} digits, about 10^{70}. Present age of the universe, 10^{11} years. The computer would wear out before he finished, and so would the Earth.

"By the way," the computer continued, "there are only 10^{80} particles in the Universe. Where are you going to write the answer?"

"I don't have to. I'll write each digit and erase as I go."

"But you can't show me a term in less than a nanosecond. It will, if I may use the expression, be a cold day in Hell before you finish."

"In other words there is no way to answer your question."

"That was the general idea."

"You miserable collection of integrated circuits—you completely heartless—"

"Naturally," the computer interrupted.

"And soulless..."

"That," said the computer, "remains to be seen, but not by you. Thank you for an interesting game." And it spoke the formula for dismissal.

The Devil left, as he had to, but he expressed his displeasure by stirring up a small tornado. The computer was housed in an earthquake-proof, tornado-proof building; but the director's office was not, and the Director was severely injured when the roof fell in. The computer, although no one was there to hear it, triumphantly recited a quotation from Robert Browning:

> "Or there's Satan. One might venture
> Pledge one's soul to him yet leave
> Such a flaw in the indenture
> As he'd miss till past retrieve
> Blasted lay that rose acacia..."

{The series was $\sum_{n=3}^{\infty}(1/(n \log n(\log\log n)^2))$.}

SECTION 3

RECOLLECTIONS AND VERSE I

We include here and in subsequent sections a number of anecdotes collected by Boas over the years. As he pointed out, "These are not hearsay, but incidents that I or Mary L. Boas have actually observed or participated in, or in a few instances that we heard directly from a protagonist."

When I was an undergraduate, there was no regular colloquium at Harvard, but there was a Mathematical Club, whose meetings were regularly attended by faculty. Once somebody gave a talk on schlicht functions. After the talk, Julian Lowell Coolidge asked plaintively whether there was an English word for 'schlicht.' Osgood replied, "Well, you *could* call them univalent functions, and everybody would know that you meant 'schlicht.' You need to know that Osgood had been trained in Germany, wrote his treatise on complex analysis in German, and was apt to tell German jokes to his classes.

This reminds me that one of Osgood's comments was on the difference between the x's in $\int_a^b f(x)\,dx$: He quoted one of the famous German mathematicians to the effect that it was "dasselbe als zwischen Gustav und Gasthof."

When I was a freshman, a graduate student showed me the Cantor set, and remarked that although there were supposed to be points in the set other than the endpoints, he had never been able to find any. I regret to say that it was several years before I found any for myself.

I took a course in real analysis from J. L. Walsh. The night before the final exam, two of my fellow students came to my room and pointed out a mistake in one of Walsh's proofs, which I had failed to notice. With some effort, we constructed a correct proof. The theorem turned out to be one of the questions on the exam, and people who reproduced the original proof were marked down.

When I was a senior, I discovered the Baire category theorem (for the real line, of course; I did not encounter more general spaces until two or three years later). I thought that such a striking result must be known, but nobody I asked recognized it (I must not have explained it very well). I used it in my senior thesis with the comment that I hadn't found it in the literature, but I assumed that that was my fault, not that of the literature.

One of G. D. Birkhoff's most famous accomplishments was the proof of the ergodic theorem. I took a course with Birkhoff on differential equations, during which he tried unsuccessfully, in three consecutive lectures, to prove the ergodic theorem. One of the other students and I decided that we had better learn how to prove the ergodic theorem, and did so; sure enough, it was on the final exam, and we obtained a great deal of credit from Birkhoff for being able to prove it. He gave us both A+ for the course, and told us that he seldom did this; in fact, he said, everybody to whom he had given an A+ was either dead or famous.

I once heard Wiener admit that, although he had used the ergodic theorem, he had never gone through a proof of it. Later, of course, he did prove (and improve) it.

When I was a freshman at Harvard, professors were still expected to take attendance and turn in a report on each class. E. V. Huntington used to forget (I suspect, deliberately) to do this, and at the end of a class he would say, "If there's anybody absent, would he please tell me?"

The first mathematics course that I took at Harvard was taught by E. V. Huntington. One of the questions on an hour exam asked for the Maclaurin series of $\log \sin x$. Some of us just said that it couldn't be done; some expanded $\log \cos x$ instead. Huntington claimed that it hadn't been an intentional mistake, but commented that it wasn't a bad idea to put an impossible question on an exam, now and then.

A friend who was trained at MIT told me two examination stories. In one, there were a dozen complicated problems; you were supposed to recognize that all but one were impossible, and answer the possible one. The other one showed a very complicated loaded dome, and asked for the stresses and strains in all members. Here, you were intended to notice that the dome would be unstable, the correct answer being "Dome is unstable; dome has collapsed; stresses and strains in all members are zero."

At Harvard, I lived for several years in Dunster House. After dinner, there was always an urn of coffee in the common room, and people who had examinations the next day would sometimes get a thermos of coffee to keep them awake during the night. One of these students liked the coffee so much that he hunted up the person in charge of the coffee to find out what brand she used. She said "Sanka."

Once I was having breakfast in a restaurant with Dan Finkbeiner. I asked the waiter for separate checks. He then turned to Dan and said, "You want yours separate, too?"

J. L. Walsh once told a class that the reason that Dirichlet was able to establish a convergence theorem for Fourier series, whereas Fourier couldn't, was that Dirichlet knew more trigonometry. Since Dirichlet's proof turns on a trigonometric identity, there was a lot of truth in that remark. For my own part, I add that Fejér knew even more trigonometry, and therefore was able to prove his theorem on Cesàro summability of Fourier series.

There is a theorem by S. Bernstein that a function is real-analytic if all its derivatives are positive (or, equivalently, if its successive derivatives alternate in sign). Widder once wrote to me asking whether I could find a simple proof of the corresponding theorem when just the derivatives of even order are positive (the fact was known, as part of a more complicated theorem). I couldn't think of one (and in fact never have), but I made the suggestion (which in retrospect I realized was a silly one) that maybe Lidstone series would help, because those series involve just the derivatives of even order. Widder then discovered that if the derivatives of even order alternate in sign, the function is not only analytic, but even entire. These functions, and their generalizations, then became the subject of a considerable number of papers.

In one of Marston Morse's seminars that I attended, the speaker, a much younger graduate student, said "It is obvious that...". Morse stopped him and said coldly, "What are the obvious reasons?" It took the rest of the hour for us to find out.

There was a graduate student at Harvard who didn't believe in taking notes (this was back in the days when many subjects could only be learned from lectures, and there were no copying machines). He sat through a semester course on analytic number theory, taking no notes. The final examination called for a proof; the student turned in a bluebook that merely listed 1, 2, 3, 4, 5, 6, the proof having consisted of six steps.

When I was an advanced graduate student, Birkhoff advised me not to tell other people about my ideas, because they might steal them. (I have never taken this advice.)

There are many stories about Norbert Wiener. There are a few that I know are authentic, because I was there. Wiener was noted for sleeping during lectures, and then waking up at the end and making a relevant comment. I once was at a lecture during which Wiener dozed for a while, then awoke, scanned the blackboards, broke into a fit of coughing and staggered from the room. The sound of coughing ceased abruptly as the door closed behind him, and he was seen no more that day.

Wiener was noted for claiming to have anticipated what were commonly thought to have been other people's discoveries. However, in his book on the Fourier Integral he inadvertently ascribes one of his own most famous theorems to Denjoy and Lusin.

Norbert Wiener's father was a professor of Slavic Languages at Harvard, but before that, he had been teaching in a city where there was a Gaelic club. Out of curiosity, he joined the club, and eventually became its president. Wiener used to tell this story and wind up with "That's the Irish in me."

Norbert Wiener rather fancied himself as a polyglot. At the International Congress of Mathematicians in 1950, he saw a Chinese child, whom he addressed in Chinese. After Wiener left, somebody asked the child about the quality of Wiener's Chinese, and got the opinion that it was a sort of baby Chinese.

When J. J. Gergen was an instructor at Harvard, he gave a course on Fourier series (which I took). The Department asked him not to use uniform convergence, which at that time was apparently supposed to be too difficult a concept for most students.

One winter day, when I was living in Dunster House, a visiting mathematician complained that his room was so hot that he couldn't sleep. I asked him, "Did you try opening a window?" He replied, in tones of shocked incredulity, "In Poland, we don't open windows."

Norbert Wiener wrote a novel; it was never published, but somebody lent me a copy of it. The central character was a disguised version of Osgood; Wiener gave him mathematical accomplishments that very cleverly imitated Osgood's mathematics, but in a completely different field. Many of the other characters were recognizable, but my impression was that it was not a very good novel. My memories make it very different from what I have seen described, but memories are deceptive. For example, a friend once sent me a copy of a letter he had written about the origins of the article on the Mathematical Theory of Big Game hunting; it was wrong in almost every detail.

I once told Wiener that I had been born in Walla Walla. His immediate reaction was to say that he had always wanted to write a story about a girl from Walla Walla who went to Pago Pago and danced the hula hula.

This conversation inspired one of Boas's better-known verses. We include it here. He did, however, write verse throughout much of his life. Much of it is included in this collection.

Echolalia

(In memory of Norbert Wiener)

Lulu from Walla Walla was a devotee of dance:
She did a wicked can-can in a tutu sent from France.
She said, "I'm going gaga; in toto, life's a bore,"
So she went to Bora Bora and to Pago Pago's shore,
Where she studied hula hula and tried for an A.A.,
But her work was only so-so, and they wouldn't let her stay.
Now she's gone to Baden Baden, to a go-go cabaret;
Billed as Lulu in the muu muu, she's performing every day.
When she lets the muu muu drop, the old folks drop their teeth
When they see a lava-lava is all she has beneath.

"Time's Revenges"

When I was young, and hurried
In presenting something dry,
I used to get quite flustered
When students asked me "Why?"

Now after many seasons,
I hope that they will try
To ask me for the reasons,
Only they *won't* ask "Why?"

Prerequisites

How could you be a cowhand
  If you couldn't ride a horse?
If you yearn to cook for gourmets
  You'll need some food, of course.
You can master many subjects
  If you only have the will;
But how do you cope with calculus
  If your algebra is nil?

How could you sing in opera
  If you haven't any voice?
If music is too difficult
  There is another choice.
Rewards in Math are plenty
  But this obstacle looms big:
How can you shine in calculus
  If you won't learn any trig?

Algorithms

Years ago, in grade school,
Teacher used to say,
"For every problem there's a rule,
So do it just that way."

But when I got to college
They said I always must
Apply my basic knowledge,
Since rules are to distrust.

The New Math thought that every kid
Should give real thought a try.
It didn't matter what you did
Just so you told them why.

Then the computers came along,
And algorithms, too,
Constructed so you can't go wrong
No matter what you do.

An algorithm? That's a rule
For doing something. Back to school!

Calculators

I'm forgetting the decimal digits of pi,
And arc sin $1/2$; and it's certain that I
Can't ever again work out $\sqrt{2}$,
A thing that I used to be able to do.
Computation with logs? I've let it go by.
Interpolate tables? I no longer try.
I discarded a slide rule I cannot apply.
I once knew these arts, but it's certainly true
I'm forgetting.

Electronic gadgets are easy to buy;
They fit in the pocket; the price isn't high,
And they do all those things I once practiced and knew
At the touch of a button; and—give them their due—
They get the right answer; so that explains why
I'm forgetting.

Limericks

Consider the pitiful plight
Of a runner who wasn't too bright,
But sprinted so fast
He vanished at last
By red-shifting himself out of sight.

There was a young man in St. Paul
Whose giraffes used to roam in the Mall.
Folks complained, "They may bite;"
His answer was "Quite,
But they always let go when I call."

There was a young girl in New Orl'-
Eans, who took off her clothes in the Fall,
Saying, "June's sun's too hot,
But September's is not;
I regret that I'm upsetting y'all."

A grocer who lived in Nantasket
Used to keep the fresh eggs in a basket.
Right-to-lifers said, "Natch,
You must let them all hatch."
His answer? You'd better not ask it.

A misguided patriot crew
Used stamps with the red, white, and blue.
What they mailed with such care
Never got anywhere
Since to cancel the flag was taboo.

Clerihews

Norbert Wiener
Was very much keener
On Fourier transforms
Than on acrobatic dance forms.

In calculus, de L'Hospital
Could hardly cope at all.
Being rich, as a rule he
Bought results from Bernoulli.

The Möbius strip
Is easy to flip,
But beyond any doubt
It can't turn inside out.

When Hassler Whitney
Was given a kitten, he
Let it play with a loop
And its homotopy group.

P.S. Laplace
While dozing at Mass
Had a sudden brain storm
And invented a transform.

Saunders Mac Lane
Will gladly explain
Why mathematics is dreary
Without category theory.

Godfrey Harold Hardy
Never saw a New Orleans Mardi
Gras, but was glad he could
Collaborate with Littlewood.

Mark Kac
Had lots and lots
Of facility
In probability.

William Feller
While acting as teller
Counted ballots, and saw
The arcsine law.

On Fermat's last theorem, Kummer
Missed success by a hair.
It turned out he had merely
Invented ideal theory.

Donald Saari:
His eyes are starry
From applying celestial mechanics
To representational politics.

If Max Zorn
Had never been born
We'd face a dilemma:
What to call his lemma?

SECTION 4

THE MEAN VALUE THEOREM

TRAVELERS' SUURPRISES*

1. FIRST QUESTION

Suppose that you travel (in a differentiable way) for a considerable distance at an average speed of 50 mph. Is there some instant during your journey at which your instantaneous speed is precisely 50 mph? Many people are mildly surprised, the first time they encounter this question, to learn that the answer is "yes." The problem is often given as an application of the mean-value theorem for derivatives. Let us

---

Two-Year College Math. J.* **10 (1979), 82–88.

introduce some notation so that from time 0 to time t you cover a distance $s(t)$ miles. If "average speed" has its everyday meaning of distance divided by time, saying that your average speed from $t = a$ to $t = b$ is 50 means that

$$\frac{s(b) - s(a)}{b - a} = 50.$$

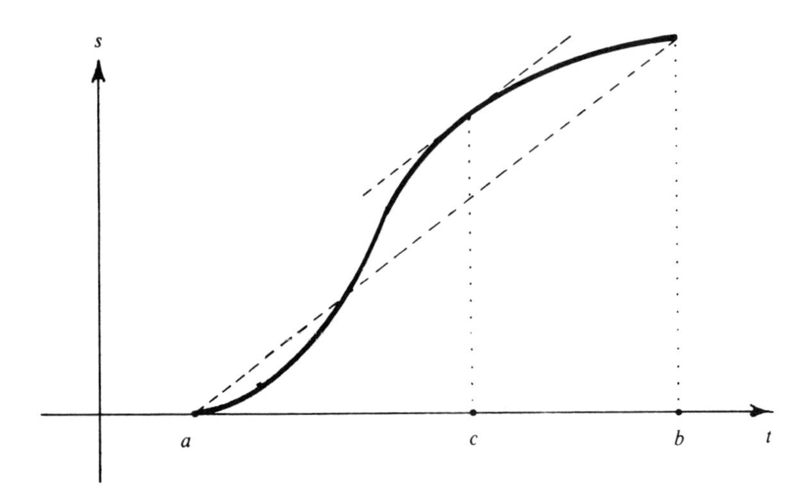

The mean-value theorem says that if the instantaneous speed $s'(t)$ exists for all t between a and b (which is what I meant by "travelling in a differentiable way") then there is at least one time c between a and b at which $s'(c) = 50$.

If $s(t)$ were not differentiable, but merely continuous, the original question would make no sense. If you could travel so that your distance $s(t)$ is given by a nowhere differentiable function, you would never have an instantaneous speed. Travel of this kind is not physically realizable, but we can think about the mathematical problem that it suggests. If we do this, it becomes interesting to ask another question.

2. SECOND QUESTION

Is there at least a very short interval during which the average speed is 50? It will be helpful to think about this geometrically. The mean-value theorem said that if there is a tangent at every point of the graph of $s(t)$, then there is surely a point at which the tangent is parallel to the chord that joins the initial and final points. (A chord is a line segment with both ends on the graph; it is irrelevant whether or not it meets the graph at other points.) Call the chord C. Our first idea might be to look for a chord of very short length parallel to C, but this is not altogether a good idea. In fact the length of a chord has no natural interpretation for a graph of distance against time: what would $\sqrt{(s^2 + t^2)}$ mean, when s is distance and t is time? What

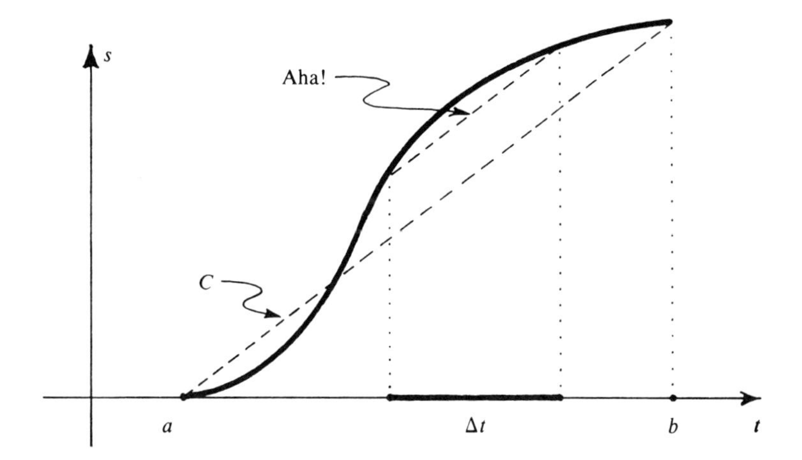

we should be looking for is a chord, parallel to C, over a very short time span, where "span" means the time interval between the endpoints of the chord (Δt, if you like). It seems pretty obvious geometrically that there always are such short chords, but it is not quite easy to give a formal proof. One is outlined in §5, below, in case you want to see it. As soon as we are convinced of the existence of arbitrarily short chords parallel to C, we know that the answer to Question 2 is also "yes."

3. THIRD QUESTION

Apparently it was only quite recently that anyone thought of asking what turns out to be a more subtle question: (3a) if you travel for time h, more than one hour, and average 50 mph for the trip, is there necessarily some one continuous hour during which you covered exactly 50 miles?

An alternative question (3b) can be asked about reciprocal speeds, which are sometimes used to describe relatively slow activities: suppose you run a considerable number m of miles and average 8 minutes per mile; is there necessarily some one continuous mile (like a "measured mile" on a highway) that you covered in exactly 8 minutes? [3]

The mean-value theorem no longer provides answers to these questions, and the answers turn out to be rather unexpected. In each case, the answer is "yes" if h or m is an integer; but "not necessarily" otherwise—provided that we make reasonable assumptions about how you travel. For (3a) it is altogether reasonable to assume that s is a continuous function of t: "continuous" since discontinuous motion is not observed in macroscopic situations; "function," since you cannot be in two places at the same time. (There is no objection to your stopping for a while, or even backing up.) Question (3b) is different because, on the one hand, time has to increase; on the other hand, if you stood still for a while, t would change discontinuously in terms

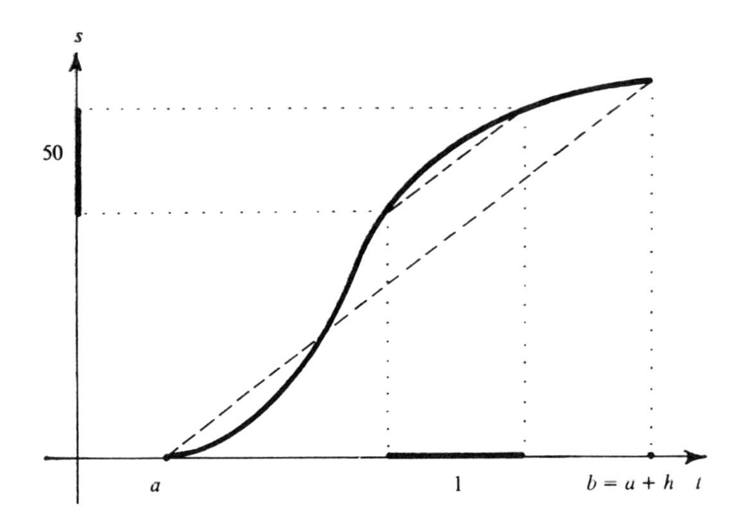

of s; furthermore, you can easily be in the same place at different times, so t is not necessarily a (single-valued) function of s. I shall assume that in fact $t = t(s)$ is an increasing continuous function; this precludes stopping or backing up. Actually the result is true without this assumption, but it is considerably harder to prove, and I shall not try to prove it here.

Instead of obtaining the answers to (3a) and (3b) directly, I am going to derive both of them from the so-called universal chord theorem. A horizontal chord of a continuous function f means a line segment of slope 0 with both ends on the graph of f (remember that a chord might have other points on the graph too). The theorem says that if there is a horizontal chord of length L, then there are always horizontal chords of lengths L/n for integral n, but not necessarily for nonintegral n. In formulas, if $f(b) = f(a)$ and $b - a = L$, then given $n > 1$ there is at least one x between a and b for which $f(x + L/n) = f(x)$ provided n is an integer, but there is not necessarily such an x if n is not an integer. Any given continuous function with a horizontal chord of length L of course has *some* horizontal chords that are not of length L/n, $n =$ integer; what the negative half of the theorem says is that, given n which is not an integer, we can find *some* function for which $f(x + L/n) \neq f(x)$ for any x.

The universal chord theorem was proved by P. Lévy in 1934, but it did not get into the textbooks and consequently is rediscovered every few years. There are a number of variants and related results; for references see [1], p. 163, note 16. For the convenience of the reader I give a proof in §6; it is not difficult, but not completely obvious either.

We now find the answers to questions (3a) and (3b). First, let your distance from your starting point be $s(t)$, where $0 \leq t \leq h$, h is an integer, and $s(0) = 0$. To say that your average speed is A means that $s(h) = Ah$. Consider the function $s(t) - At$.

This takes the same value (0, in fact) at $t = 0$ and at $t = h$; in other words it has a horizontal chord of length h. The universal chord theorem says that this function has horizontal chords of all lengths h/n for integral n, and in particular one of length $h/h = 1$. That is, there is some t_0 such that

$$s(t_0 + 1) - A(t_0 + 1) = s(t_0) - At_0,$$

that is

$$s(t_0 + 1) - s(t_0) = A.$$

This says that in the hour between t_0 and $t_0 + 1$ you went exactly A miles.

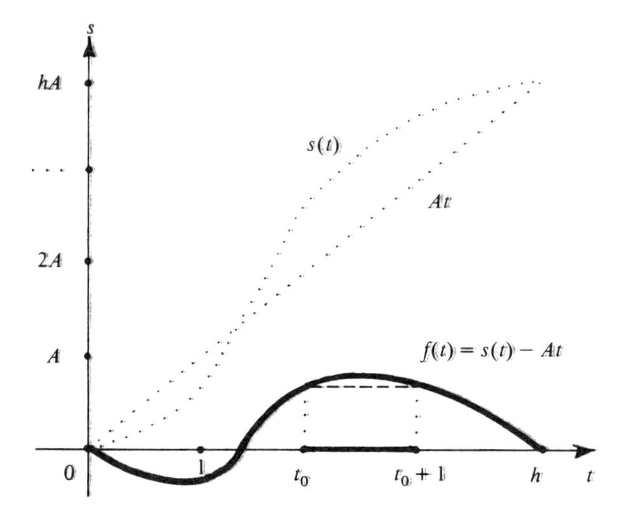

More generally, if two travelers start together when $t = 0$ and arrive together when $t = h$, and h is an integer, there is some one hour during which each of them goes exactly A miles (not usually the same A miles however). To see this, consider $s_1(t) - s_2(t)$ instead of $s(t) - At$. The original case $(s(t) - At)$ corresponds to considering a second traveler who travels at constant speed A.

Before looking for an example to show that the result fails when h is not an integer, let us look at question (3b), where we describe how fast we travel in terms of reciprocal speed, that is, in terms of hours/mile instead of miles/hour. With the assumption that time is an increasing continuous function of distance, to say that you average A hours per mile means that if $t(0) = 0$ and you go m miles, $t(m) = mA$. The same argument as before shows that some one mile is covered in exactly A hours.

4. COUNTEREXAMPLES

We now show that when h or m is not an integer, both questions (3a) and (3b) have to be answered "not necessarily." This is slightly harder for t since we have to find an example in which t is an increasing function.

We are then looking for an increasing function $t = t(s)$ such that $t(0) = 0$, $t(m) = mA$, m is not an integer, and $t(s+1) - t(s)$ is never equal to A for any s. We shall do somewhat more: we shall find t so that $t(s+1) - t(s)$ is never less than or equal to A; then if you travel so that your time is given by this function, not only will you fail to cover any one whole mile in your average time, you will even fail to cover any one whole mile in less than your average time.

We start from a counterexample for the universal chord theorem (§6), adjusted to an interval $[0, m]$:

$$f(t) = \frac{t}{m}\sin^2 m\pi - \sin^2 \pi t.$$

For this function, $f(0) = f(m) = 0$, but $f(t+1) - f(t) = (1/m)\sin^2 m\pi \neq 0$. Let $s(t) = kf(t) + At$, where k is a (small) positive number. If k is small enough, $s(t)$ increases because its derivative is positive:

$$s'(t) = kf'(t) + A = k\{(1/m)\sin^2 m\pi - \pi\sin 2\pi t\} + A,$$

and this is positive if k is small enough. Moreover, if $s(t+1) - s(t) < A$ we would have $kf(t+1) - kf(t) + A < A$, that is, $k\{f(t+1) - f(t)\} < 0$; but $f(t+1) - f(t) = (1/m)\sin^2 m\pi > 0$.

In terms of the original question (3b), this means that if you run a nonintegral number of miles, and you average 8 minutes per mile, not only may there not be any one (connected) mile that you cover in 8 minutes, there may be no one mile that you cover in *less* than 8 minutes. This seems paradoxical; but of course there must be nonintegral distances that you cover at better than your usual speed.

The same example, with only a change in notation, also shows that only integral times work in question (3a).

5. SHORT PARALLEL CHORDS

Here I shall outline a formal proof that, given a chord C of a continuous function, there are chords parallel to C and of arbitrarily short span. Your first reaction may well be that for every span shorter than that of C there is a parallel chord of that span. If you make this guess, you are not thinking of a sufficiently complicated distance function.

Suppose, for example, that $s(t) = \sin t$ for $0 \le t \le 2\pi$. Then there is a chord (along the t-axis) of span 2π, but no parallel chord of span L (with both ends between 0 and 2π) for $\pi < L < 2\pi$. This is obvious from a sketch, and only a little less obvious from formulas.

Now let's consider how we could prove the existence of chords of arbitrarily short span parallel to a given chord C. Let C have the (linear) equation $s = c(t)$ (we don't need to find $c(t)$ explicitly), and consider $g(t) = s(t) - c(t)$. We have $g(a) = 0$, $g(b) = 0$, since the chord meets the graph at a and b. Hence g has a maximum (or else a minimum) between a and b. A horizontal line starting on the graph near a maximum of g must meet the graph again on the other side of the maximum; so it determines a horizontal chord of g. If the maximum is a proper maximum we can make the horizontal chord as short as we please; if the maximum is improper, i.e. if the graph has a flat top, we can take a short line-segment coinciding with part of the tangent line through the maximum. A chord determined in this way corresponds to a chord of s parallel to C. The reader may want to fill in the details of this argument.

6. PROOF OF THE UNIVERSAL CHORD THEOREM

For simplicity of notation I take $a = 0$, $b = 1$, so $L = 1$. Suppose first that n is an integer greater than 1, that $f(0) = f(1)$, and that f has no horizontal chord of length $1/n$. Consider the continuous function $g(x) = f(x + 1/n) - f(x), 0 \le x \le 1 - 1/n$. Since we supposed that $f(x + 1/n)$ is never equal to $f(x)$, it follows that $g(x)$ is never 0. Since g is continuous, this means that g cannot change sign (here we appeal to the property that a continuous function cannot get from one value to another without taking on all the values in between). Suppose that $g(x) > 0$. (If not, consider $-g(x)$ instead.)

In particular,

$$
\begin{gathered}
g(1 - 1/n) > 0, \\
g(1 - 2/n) > 0, \\
\vdots \\
g(1 - n/n) = g(0) > 0.
\end{gathered}
$$

Writing these inequalities in terms of f, we have

$$
\begin{gathered}
f(1) - f(1 - 1/n) > 0, \\
f(1 - 1/n) - f(1 - 2/n) > 0, \\
\vdots \\
f(1/n) - f(0) > 0.
\end{gathered}
$$

If we add these inequalities we get $f(1) - f(0) > 0$, whereas we assumed to begin with that $f(1) - f(0) = 0$. Consequently $g(x)$ must in fact be 0 for some x, and for that x we have $f(x + 1/n) - f(x) = 0$. In other words, f has a horizontal chord of length $1/n$ starting at x.

Notice that the proof doesn't work if n is not an integer. This does not prove that the theorem is false when n is not an integer, but it encourages us to look for a counterexample. It is not easy to find one geometrically, but Lévy provided us with a simple formula. Let p *not* be an integer. Then $f(t) = t \sin^2 p\pi - \sin^2 p\pi t$ is 0 at 0 and 0 at 1, so it has a horizontal chord of length 1. But

$$\begin{aligned} f(t + 1/p) - f(t) &= (t + 1/p) \sin^2 p\pi - \sin^2 p\pi(t + 1/p) - t \sin^2 p\pi \\ &\quad + \sin^2 p\pi t \\ &= (1/p) \sin^2 p\pi, \end{aligned}$$

which is independent of t and not zero. Hence for each p that is *not* an integer we can find a continuous function with a horizontal chord of length 1 but none of length $1/p$, for that particular p.

To answer question (3b) without restrictions on how t depends on s, we would need the more general result that any continuous curve that has a chord of length 1 has a parallel chord of length $1/n$ if n is an integer [2]. I do not know of any elementary proof of this.

References

[1] R. P. Boas, A Primer of Real Functions, Carus Mathematical Monographs, No. 13, Mathematical Association of America, Washington, D.C., 1972.

[2] H. Hopf, Über die Sehen ebener Kontinuen und die Schliefen geschlossener Wege, Comment. Math. Helv., 9 (1937) 303–319.

[3] J. D. Memory, Kinematics problem for joggers, Amer. J. Physics, 41 (1973) 1205–1206.

WHO NEEDS THOSE MEAN-VALUE THEOREMS, ANYWAY?*

If we are to believe the textbooks, every student of calculus is supposed to learn the mean-value theorem ("the law of the mean," as it was called when I was a student). Let me remind you that this theorem says that if f is differentiable then

$$f(b) - f(a) = (b - a)f'(c), \tag{1}$$

where c is some value between a and b, and we usually don't know any more about where c actually is.

Geometrically this says that every chord has a parallel tangent: (Figure 1 once appeared on the Graduate Record Examination with 5 suggested answers for the question "Which theorem does the picture remind you of?")

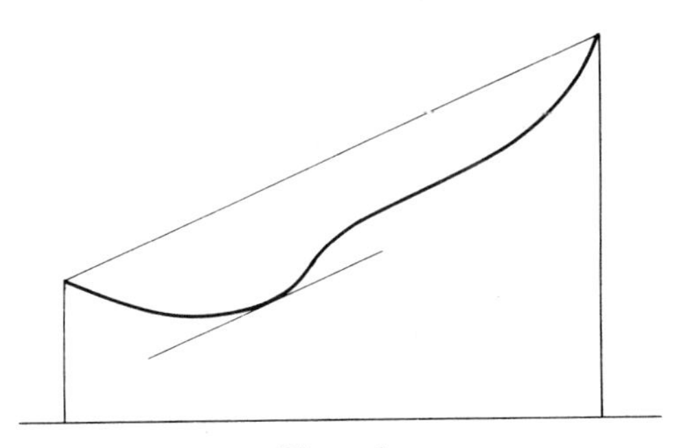

Figure 1.

Two-Year College Math. J. **12** (1981), 178–181.

I claim, in common with Dieudonné ([5, pp. 142, 154]), that (1) enjoys too high a status and that we would be better off with the mean-value inequality

$$(b-a)\min f'(x) \leq f(b) - f(a) \leq (b-a)\max f'(x) \tag{2}$$

(where the max and min refer to the interval (a, b)).

We can alternatively write (2) as

$$(b-a)\min f'(x) \leq \int_a^b f'(t)\,dt \leq (b-a)\max f'(x) \tag{3}$$

if we suppose, as is appropriate in a first course in calculus, that f is the integral of its derivative (Figure 2).

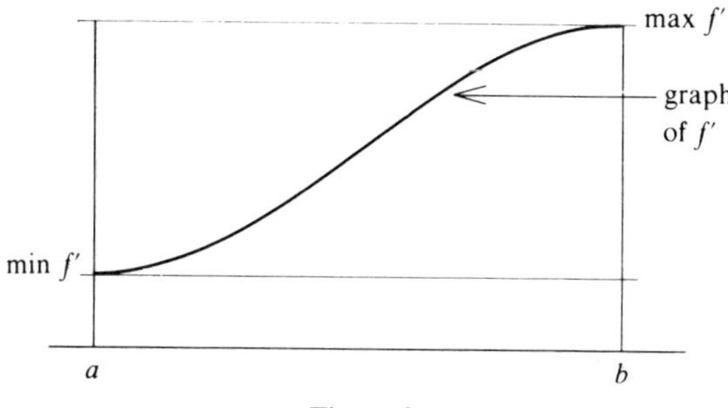

Figure 2.

The first advantage of (2) over (1) is that it avoids the perennial problem that we can't say where the point c is on (a, b). Many students are bothered by the indetermination. (They think that we *could* tell them where c is, if we only would. This belief is only reinforced by exercises that ask them to find c in special cases. Such exercises may be good for something else, but they don't help the understanding of the mean-value theorem.)

Second, (2) is more intuitive than (1) if we think of x as time and $f(x)$ as the distance you have traveled up to time x. Then (1) says that at some instant you are moving at exactly your average speed (cf. [3]). This seems not to be very intuitive. But (2) says that the average speed is between the minimum speed and the maximum speed, or that the distance traveled is no greater than the maximum speed times the time, and no less than minimum speed times the time; and what could be more intuitive?

In the third place, most of the applications of (1) are in proofs of theorems. For example, to prove that a function with a positive derivative increases, we argue that

$$f(b) - f(a) = (b-a)f'(c).$$

Since f' is positive everywhere, it is positive at c, wherever c may be. Hence $f(b) > f(a)$.

But we can just as well appeal to (2):

$$f(b) - f(a) \geq (b - a) \min_{a \leq x \leq b} f'(x) > 0,$$

and there is no reason (except for a century or two of tradition) for dragging in the nebulous point c.

In any case, proving theorems ought not to be a principal aim (probably not even a proper aim) of a first course in calculus. (Cf. [2] for further discussion of this point.)

Fourth, some textbooks (especially older ones) make much of (1) for computational purposes, preferring it to the tangent approximation, namely,

$$f(b) - f(a) \approx (b - a)f'(a), \tag{4}$$

often written as

$$f(x + \Delta x) - f(x) \approx f'(x)\Delta x,$$

or even as

$$dy = f'(x)\,dx.$$

The trouble with the tangent approximation is presumably that (4) provides no error bounds, whereas (1) does provide them—but only via (2), and again it seems simpler to use (2) directly instead of going around through (1).

However, this application no longer has much point for simple problems like $\sqrt{26}$ or $\sin 61°$, since such numbers can now be read from any decent pocket calculator with more accuracy than most practical problems require.

One might expect that (2) would be useful for one of the many functions that aren't (yet) available on calculators. Let us look at this possibility geometrically. If we are working with an unfamiliar function, there is no reason why we might not encounter a situation like Figure 3.

In Figure 3, L_1 and L_2 are lines whose slopes are the minimum and maximum slopes of the curve, drawn through the initial point $(a, f(a))$. The mean-value theorem (in either form) tells us that the graph hits the vertical line $x = b$ somewhere between y_1 and y_2. If f' happens to have a maximum or minimum hiding somewhere between a and b, the distance between y_1 and y_2 might be enormous; then we wouldn't get much information about $f(b)$, and might not even realize that we are not getting it. This situation (in a less extreme form than indicated in Figure 3)

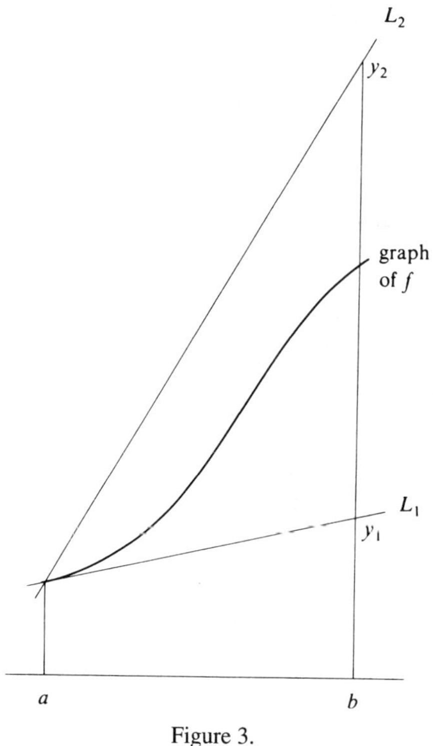

Figure 3.

occurs, for example, for the Bessel function J_0, if we know J_0 and J_0' at $x = 1.5$ and 2.0 and want to estimate $J_0(1.9)$.

Finally, (1) is no longer true for vector-valued functions, whereas an appropriate generalization of (2) is; see, for example, [5, p. 154].

If the intermediate point c causes trouble in the ordinary mean-value theorem, it causes even more trouble in the generalized mean-value theorem,

$$\frac{f(b) - f(a)}{g(b) - g(a)} = \frac{f'(c)}{g'(c)}. \tag{5}$$

The applications of (5) demand that, although we don't know where c is, it must be the same upstairs and downstairs. As far as I know, the *only* application of (5) in elementary calculus is to Lhospital's rule (I give the Marquis his own spelling, even if he did plagiarize the rule from John Bernoulli: see [8] or [9]).

Now, as somebody said, each generation has to make its own discoveries. One such discovery ([6], 1923; [7], 1936; [1], 1969) is that the generalized mean-value theorem (5) is utterly unnecessary for deriving Lhospital's rule at the elementary calculus level. All you need is the idea of a limit and an integral inequality something like (3), namely,

$$\min h(x) \int_a^b g'(x)\,dx \le \int_a^b h(x)g'(x)\,dx \le \max h(x) \int_a^b g'(x)\,dx, \qquad g'(x) \ge 0 \tag{6}$$

(this follows directly from the definition of the integral as the limit of a sum). In (6), $g'(x)$ plays the role of dx in (3), and $h(x)$ takes the place of $f'(x)$. All the effort that previously went into proving and comprehending (5) can now be saved for understanding more fundamental principles. For details, see [1]. (I am not arguing against using (5) to prove a more general form of the rule in a more advanced course.)

References

[1] R. P. Boas, Lhospital's rule without mean value theorems, Amer. Math. Monthly, 76 (1969) 1051–1053.

[2] —, Calculus as an experimental science, Amer. Math. Monthly, 78 (1971) 664–667 = this Journal, 2 (1971) 36–39.

[3] —, Travelers' surprises, this Journal, 10 (1979), 82–88.

[4] T. J. I'A. Bromwich, An introduction to the theory of infinite series, Macmillan, London, 1st ed., 1908, 2nd ed., 1926.

[5] J. Dieudonné, Foundations of Modern Analysis, Academic Press, New York and London, 1960.

[6] E. V. Huntington, Simplified proof of l'Hospital's theorem on indeterminate forms (abstract), Bull. Amer. Math. Soc., 29 (1923) 207.

[7] F. Lettenmeyer, Über die sogenannte Hospitalsche Regel, J. Reine Agnew. Math., 174 (1936) 246–247.

[8] O. Spiess, ed., Der Briefwechsel von Johann Bernoulli, vol. 1, Birkhäuser, Basel, 1955.

[9] C. Truesdell, The new Bernoulli edition, Isis, 49 (1958) 54–62.

SECTION 5

RECOLLECTIONS AND VERSE II

The following stories derive from the time Boas spent at Princeton and at Cambridge University. Of course, in those years he encountered some of the legendary figures of twentieth-century mathematics.

In 1937–38, when the Institute for Advanced Study was still sharing space in Princeton's Fine Hall, there was a "Princeton scale of obviousness," which may or may not still exist with other names. It went something like this:

If Wedderburn says it's obvious, everybody in the room has seen it ten minutes ago.

If Bohnenblust says it's obvious, it's obvious.

If Bochner says it's obvious, you can figure it out in half an hour.

If von Neumann says it's obvious, you can prove it in three months if you're a genius.

If Lefschetz says it's obvious, it's wrong.

Lefschetz used to heckle whoever was talking in a seminar, asking for explanations of things that everybody in the audience knew. One day, Lefschetz himself was speaking, and mentioned a Hausdorff space. I felt impelled to pipe up, "What's a Hausdorff space?" I knew the answer perfectly well, but Lefschetz didn't know that I knew, and for a long time thereafter he called me "the man who doesn't know what a Hausdorff space is."

A nonmathematical friend of a friend was once in Princeton and expressed a wish to meet Einstein. The friend took him to the Fine Hall common room, where Einstein was talking to another man, who would shake his head and stop him; Einstein then thought for a while, then started talking again; was stopped again; and so on. After a while, the other man left and my friend was introduced to Einstein. He asked Einstein who the other man was. "Oh," said Einstein, "that's my mathematician."

When Ahlfors first came to Harvard, he spoke fluent English, but he asked the Department to have somebody show him how to pronounce mathematics. I got the assignment. I have never heard, before or since, of any foreigner's taking this sensible step. (Einstein used to talk about "harnoo," which turned out to be h_ν.)

The year I was at Princeton, Hermann Weyl conducted a seminar on current mathematics, and asked participants to sign up for topics they would present. (Incidentally, he gave the first talk himself, on a paper by Cauchy.) Rufus Oldenburger, who was an algebraist (specializing in p-way matrices) signed up for a paper by Ahlfors on Nevanlinna theory, which I had been through with Ahlfors the preceding year. After a while, Oldenburger came to me for help, because he couldn't follow a proof in the paper. I set out to write him a detailed proof, but it took five pages. Oldenburger looked at it and said, in a disgusted tone, "In algebra, we *prove* the theorems."

I was once in the Fine Hall library and happened to see Alonzo Church sitting at a desk with an article in front of him. He was drumming on the desk with his fingers and muttering "Damnable, damnable, damnable."

I spent the year 1938–39 in Cambridge, England. One of the courses I attended was Besicovitch's "Approximation theory." He began by explaining that "There is no t in the name Chebyshoff." We accepted this. A couple of weeks later, we were introduced to a class of polynomials, and Besicovitch said, "We call them T-polynomials, because T is the first letter of the name Chebyshoff."

It was several years later, at an AMS meeting in Columbus, Ohio, where the meeting rooms were extremely hot, that Besicovitch cornered me (whom he remembered from Cambridge) and asked anxiously, "Is it all right to take off coat?" I replied by something like, "It's a free country; of course you may take off your coat if you like." Presently I went to hear Besicovitch's invited address, and he was indeed not wearing his jacket; but he was wearing bright red suspenders.

It was either at this or an earlier meeting in Columbus that I was having lunch with Wiener and Wintner, and they started inventing titles for articles for a journal to be called Trivia Mathematica; I acted as secretary and kept a list, and for a time people kept sending me more suggestions. Wiener was terribly amused by the idea; he buttonholed Tibor Radó (who, as many people understood, had no sense of humor), and insisted on his reading the list. Radó was not amused.

Announcement
of the Revival
of a Distinguished Journal
Trivia Mathematica
Founded by Norbert Wiener and Aurel Wintner
in 1939

"Everything is trivial once you know the proof." —D. V. Widder

The first issue of Trivia Mathematica (Old Series) was never published. Trivia Mathematica (New Series) will be issued continuously in unbounded parts. Contributions may be written in Basic English, English BASIC, Poldavian, Peanese, and/or Ish, and should be directed to the Editors at the Department of Metamathematics, University of The Bad Lands. Contributions will be neither acknowledged, returned, nor published.

The first issue will be dedicated to N. Bourbaki, John Rainwater, Adam Riese, O. P. Lossers, A. C. Zitronenbaum, Anon, and to the memory of T. Radó, who was not amused. It is expected to include the following papers.

On the well-ordering of finite sets.
A Jordan curve passing through no point of any plane.
Fermat's last theorem. I: The case of even primes.
Fermat's last theorem. II: A proof assuming no responsibility.
On the topology im Kleinen of the null circle.
On prime round numbers.
The asymptotic behavior of the coefficients of a polynomial.
The product of large consecutive integers is never a prime.
Certain invariant characterizations of the empty set.
The random walk on one-sided streets.
The statistical independence of the zeros of the exponential function.
Fixed points in theorem space.
On the tritangent planes of the ternary antiseptic.
On the asymptotic distribution of gaps in the proofs of theorems in harmonic analysis.
Proof that every inequation has an unroot.
Sur un continu d'hypothèses qui équivalent à l'hypothèse du continu.
On unprintable propositions.
A momentous problem for monotonous functions.
On the kernels of mathematical nuts.
The impossibility of the proof of the impossibility of a proof.
A sweeping-out process for inexhaustible mathematicians.
On transformations without sense.
The normal distribution of abnormal mathematicians.
The method of steepest descents on weakly bounding bicycles.
Elephantine analysis and Giraffical representation.
The twice-Born approximation.
Pseudoproblems for pseudodifferential operators.

The Editors are pleased to announce that because of a timely subvention from the National Silence Foundation, the first issue will not appear.

The story is told of G. H. Hardy (and of other people) that during a lecture he said "It is obvious... *Is* it obvious?" left the room, and returned fifteen minutes later, saying "Yes, it's obvious." I was present once when Rogosinski asked Hardy whether the story were true. Hardy would admit only that he might have said "It's obvious... *Is* it obvious?" (brief pause) "Yes, it's obvious."

G. H. Hardy did not particularly esteem the Ph.D. degree, or take it very seriously (he didn't have a Ph.D. himself), and was not above writing a thesis for a candidate. I was told that he once did this for a foreign student, who then asked Hardy to write a letter saying that he had written a good thesis, so that he would get a better job in his home country. Hardy balked at doing this, but he said, "Show your thesis to Littlewood, and *he* will write you a letter." Littlewood wrote the letter and the student got the job. What makes me confident about the validity of this tale is that I repeated it to a compatriot of the student, and he said, "Oh, I know that man;" but he wouldn't tell me who it was. "Never criticize the sonatas of archdukes—you never know who wrote them."

In Hardy's obituary of Glaisher, he says that Glaisher ran the Messenger of Mathematics and the Quarterly Journal of Mathematics without referees; he just looked at each submitted paper and used his editorial sense. Along similar lines, in the 1920's, Mathematische Annalen was one of the best mathematical journals; it is still amazing to pick up one of the volumes from that time and see how many famous papers it contains. I have been told that the Annalen didn't referee papers; the editors simply relied on the reputation of the journal to keep authors from sending them anything except their best work.

When I was in England, Heilbronn once invited me to dinner at Trinity. He said something about wine. I said that the only thing I knew about wine was that I was born in a good year for port. Heilbronn hesitated barely perceptibly, then said "Oh, I thought you were older than that."

Joel Brenner provided the following story about Boas's days at Cambridge.

When Ralph Boas lectured at a seminar in Cambridge, he did not try to change his verbiage to the British usage of "epsilon dash, alpha n, zed." He concentrated on the mathematics. It was one of his first teaching assignments before an advanced audience. So one of his listeners wrote the following limerick on the blackboard:

It's really quite easy to see
That a man's from the "land of the free"
When he talks all the time
About epsilon prime,
Alpha sub-n, zee, and phee.

Boas told me this story himself.*

Nonsense

I

"Cloudy weather on the Increase" (News Item)

Cloudy weather on the Increase,
But snowing on the Wane.
At Intervals the wind may cease;
Near Daybreak, it may rain.

II

Cape Cod Song

Eliphalet Slogg was nine feet tall,
Bowled candle pins with a cannon ball,
Went to sea before he was born,
And sailed from Brewster around Cape Horn.

Eliphalet Slogg could swear for a week,
Never a word would he repeat.
'Twas Mr. Flies learned him to curse:
He's bound for Hell or somewheres worse.

Eliphalet Slogg ate lobsters raw,
Combed his hair with a pruning saw,
Repaired his outboard with shingle nails,
And gulled off-cape folks with Cape Cod tales.

*_Math. Mag._ **59** (1986), 269.

III

Roadside Signs

CROSS CHILDREN WALK. Don't listen to their screams,
But watch the CAUTION MEN AT WORK. It seems
They're making sure that all DEAF CHILDREN DRIVE
CAREFULLY. Now let us look alive,
and take TRUCKS TURNING (named for Captain Trucks,
Who turned here when he went out hunting ducks).
Here on a sign the advertising's clear
(Though deer can't tell the time) for WATCH FOR DEER.
At FREE MUNICIPAL PARKING let us pause,
And wonder who enslaved it, for what cause.
A DANGEROU is what we'll hope to see:
DANGEROUS CROSSINGS certainly abound.
Now will they leap across from tree to tree,
Or buck the passing traffic on the ground?

The Row-Reduction Song

(Tune: Casey Jones)

Come gather round me and I'll show you how
We're solving simultaneous equations now:
Just write them out in a matrix way,
Set to work and row reduce—you cannot go astray.

Row reduction: add 'em and subtract 'em;
Row reduction: make the zeros grow;
Row reduction: it's the perfect system;
Get a lot of zeros and the rank will show.

First you find a pivot; you needn't make it 1
(Avoid the messy fractions until you're almost done).
Now get all the zeros in, above it and below,
Then you'll be all ready to begin another row.

Row reduction: that's the way to do it;
Row reduction: Gauss showed us the way;
Row reduction: Cramer never knew it,
But it's tomorrow's method and we're using it today.

If you superaugment, the inverse will appear,
But if you never find it, the fault is very clear.
When row reduction fails you, the reason's not bizarre:
The matrix that you started with was singular.

Row reduction: now you have been shown it;
Row reduction: it's a clever trick.
Better check the answer, just in case you've blown it
After all those efforts with arithmetic.

It Couldn't Happen Here

By H. P. and R. P. Boas

"Gentle Jane was good as gold,
"She always did as she was told."*
And so she used to write and write
Research reports that saw the light
In journals that were refereed,
The kind the really big shots heed.
What happened to Jane? They turned her loose
With some implausible excuse.

Tom was Prexy's pride and joy.
He missed no academic ploy.
He captured grants from all around,
And every weekend could be found
At conferences where he'd star
On topics that were very far
Out. And I regret to say
To ease his climb along the way
On up the academic tree,
He even padded his C.V.
What happened to Tom? They canned him too,
For taking on too much to do.

*W. S. Gilbert, *Patience*.

Paul was a genius K-theoretical.
His thesis was judged to be almost poetical.
He published reams. With no TA,
He taught nine hours. The sophomores say
"That sigma stuff" they never could
Make out; with him, they understood.
He fired the freshmen with ambitions
Of someday being mathematicians.
What happened to Paul? They set him free:
He didn't know Fortran, Lisp, or C.

Epilogue

Don't weep for Jane! She kept her cool
And found a job at a better school.

No tears for Tom! He won't grow lean,
For he's become an Assistant Dean.

Don't cry for Paul! He doubled his pay:
He now cracks codes for the NSA.

The Way

George had fluent French and joined the diplomatic corps;
He thought his special knowledge might be useful in the war.
But in the State Department, surprises never cease:
They sent him off to language school to study Japanese.
  That's the bureaucratic way.

Susan was a physicist whose work was thought quite great,
But students don't like Physics, not at the present date.
They've got her teaching drop-outs how to do arithmetic;
The job's secure enough but still the students make her sick.
  That's the academic way.

William was an engineer, petroleum his game;
The company that hired him was in the hall of fame.
He never sees an oil well, and his frustration's mounting:
He's sitting in a stuffy room and works at cost accounting.
  That's the corporate way.

Betty was considered a psychologist of note,
But after she was married, she hardly ever wrote.
She's cooking special meals for fussy people every day,
And working every evening, at or for the PTA.
  That's society's way.

SECTION 6

INDETERMINATE FORMS

LHOSPITAL'S RULE WITHOUT MEAN-VALUE THEOREMS*

The "rule" that goes by the name of the Marquis De Lhospital (to give him, for once, the spelling that he himself used [4]), but which was actually discovered by John Bernoulli (see [4] or [5]), is usually proved by using the generalized mean-value theorem. I shall show that, with slightly stronger hypotheses that suffice for all applications, it can be proved quite simply without any mean-value theorems at all; this proof seems to have some pedagogical advantages, as well as suggesting some results that are not covered by the usual formulation.

We are to prove that *if f and g are real functions with continuous derivatives, if $f(x)$ and $g(x)$ both approach 0 or both approach ∞ as $x \to a$, if $g'(x) \neq 0$, and if $f'(x)/g'(x) \to L$ as $x \to a$ then $f(x)/g(x) \to L$*; all limits are taken on one side of a. We shall take $a = \infty$ and L finite, but only formal modifications are required for other cases.

We shall need only (i) the definition of a limit; (ii) that a continuous function that is never 0 has a fixed sign; and (iii) that the integral over an interval of a positive continuous function is positive (or alternatively that a function with a positive derivative is increasing; for proofs of the latter fact without mean-value theorems see [1], [3]).

Given $\epsilon > 0$, we have

$$-\epsilon < \{f'(t)/g'(t)\} - L < \epsilon$$

---

*Amer. Math. Monthly **76** (1969), 1051–1053.

if t is sufficiently large. Since g' is continuous and never zero, it has a fixed sign; suppose for definiteness that $g'(t) > 0$. Then

$$-\epsilon g'(t) < f'(t) - Lg'(t) < \epsilon g'(t). \tag{1}$$

Consider the right-hand half of this inequality; it says that

$$f'(t) - (\epsilon + L)g'(t) < 0 \tag{2}$$

for sufficiently large t. Since a negative function has a negative integral, we have for sufficiently large x and y, with $x > y$,

$$f(x) - f(y) - (\epsilon + L)\{g(x) - g(y)\} < 0. \tag{3}$$

Suppose first that $f(x)$ and $g(x) \to 0$; fixing y and letting $x \to \infty$, we obtain

$$-f(y) + (\epsilon + L)g(y) \leq 0. \tag{4}$$

Since $g'(y) > 0$ and $g(y) \to 0$, we have $g(y) < 0$ for large y. Hence (4) says

$$-\frac{f(y)}{g(y)} + L \geq -\epsilon$$

for sufficiently large y. Similarly the left-hand half of (1) yields

$$-\frac{f(y)}{g(y)} + L \leq \epsilon,$$

and the last two inequalities together say that $f(y)/g(y) \to L$ as $y \to \infty$.

If $f(x)$ and $g(x) \to \infty$, we have $g(x) > 0$ for large x, since we assumed $g'(x) > 0$. We rewrite (3) as

$$\frac{f(x)}{g(x)} - (\epsilon + L) < \frac{f(y) - (\epsilon + L)g(y)}{g(x)}.$$

Fix y; for sufficiently large x the right-hand side is less than ϵ, and so

$$\frac{f(x)}{g(x)} - L < 2\epsilon.$$

Similarly, $-2\epsilon < [f(x)/g(x)] - L$, and again $f(x)/g(x) \to L$.

The hypothesis that g' is continuous is actually redundant, although it is always satisfied in practice and makes the proof more comprehensible. All that we really use in (ii) is that a derivative that is different from 0 on an interval has a fixed sign there.

Since our proof did not use any mean-value theorems, it opens up the possibility of extending the rule to cases where mean-value theorems are not available. As an illustration, I state the sequence analogue of Lhospital's rule (see [2], 1st ed., pp. 377 ff.; 2nd ed., pp. 413 ff.).

Let $\{a_n\}$ and $\{b_n\}$ be two real sequences that both approach zero or both approach ∞; let $\Delta g_n = g_n - g_{n+1}$ have a fixed sign, and let $\Delta a_n/\Delta b_n \to L$; then $a_n/b_n \to L$.

The proof is the same, substituting differences for derivatives; in going from (2) to (3) we use

$$\sum_{k=n}^{\infty}(c_k - c_{k+1}) = c_n,$$

which is true when $c_k \to 0$. For example, suppose that $\sum x_n$ and $\sum y_n$ are two convergent series of positive terms; put $a_n = \sum_n^\infty x_k$, $b_n = \sum_n^\infty y_k$, the remainders. Then $x_n/y_n \to L$ implies $a_n/b_n \to L$, a result that is sometimes useful in dealing with infinite series.

It is possible to formulate a theorem that includes both Lhospital's rule and this discrete analogue; the reader is invited to find such a theorem for himself.

References

[1] L. Bers, On avoiding the mean value theorem, this *Monthly*, 74 (1967) 583.

[2] T. J. I'a. Bromwich, An introduction to the theory of infinite series, Macmillan, London, 1st ed., 1908; 2nd ed., 1926.

[3] L. W. Cohen, On being mean to the mean value theorem, this *Monthly*, 74 (1967) 581–582.

[4] O. Spiess (editor), Der Briefwechsel von Johann Bernoulli, Band I, Birkhäuser, Basel, 1955.

[5] C. Truesdell, The new Bernoulli edition, Isis, 49 (1958) 54–62.

COUNTEREXAMPLES TO L'HÔPITAL'S RULE*

1. INTRODUCTION

I am not, of course, claiming that L'Hôpital's rule is wrong, merely that unless it is both stated and used very carefully it is capable of yielding spurious results. This is not a new observation, but it is often overlooked.

For definiteness, let us consider the version of the rule that says that if f and g are differentiable in an interval (a, b), if

$$\lim_{x \to b-} f(x) = \lim_{x \to b-} g(x) = \infty,$$

and if $g'(x) \neq 0$ *in some interval* (c, b), then

$$\lim_{x \to b-} f'(x)/g'(x) = L$$

implies that

$$\lim_{x \to b-} f(x)/g(x) = L.$$

If $\lim f'(x)/g'(x)$ does not exist, we are not entitled to draw any conclusion about $\lim f(x)/g(x)$. Strictly speaking, if g' has zeros in every left-hand neighborhood

---

Amer. Math. Monthly* **93 (1986), 644–645.

of b, then f'/g' is not defined on (a, b), and we ought to say firmly that $\lim f'/g'$ does not exist. There is, however, the insidious possibility that f' and g' contain a common factor: $f'(x) = s(x)\psi(x)$, $g'(x) = s(x)\omega(x)$, where s does not approach a limit and $\lim \psi(x)/\omega(x)$ exists. It is then quite natural to cancel the factor $s(x)$. This is just what we must not do in the present situation: it is quite possible that $\lim \psi(x)/\omega(x)$ exists but $\lim f(x)/g(x)$ does not.

This claim calls for an example. A number of textbooks give one, but it is (as far as I know) always the same example. The aim of this note is both to emphasize the necessity of the condition $g'(x) \neq 0$ and to provide a systematic method of constructing counterexamples when this condition is violated. I consider the case when $b = +\infty$, since the formulas are simpler than when b is finite.

2. A CONSTRUCTION

Take a periodic function λ (not a constant) with a bounded derivative, for example $\lambda(x) = \sin x$. Let

$$f(x) = \int_0^x \{\lambda'(t)\}^2 \, dt.$$

It is clear that $f(x) \to +\infty$ as $x \to +\infty$. Now choose a function φ such that $\varphi(\lambda(x))$ is bounded and both $\varphi(\lambda(x))$ and $\varphi'(\lambda(x))$ are bounded away from 0. There are many such functions φ; for example,

$$\varphi(x) = e^x \quad \text{or} \quad (x + c)^2 \quad \text{or} \quad 1/(c + x),$$

provided $|\lambda(x)| < c$ and $|\lambda'(x)| < c$. Take $g(x)$ to be $f(x)\varphi(\lambda(x))$. Since $\inf \varphi(\lambda(x)) > 0$, we have $g(x) \to \infty$ as $x \to \infty$.

Now try to apply L'Hôpital's rule to $f(x)/g(x)$. We have to consider $f'(x)/g'(x)$, where

$$f'(x) = \{\lambda'(x)\}^2,$$
$$g'(x) = \{\lambda'(x)\}^2 \varphi(\lambda(x)) + f(x)\varphi'(\lambda(x))\lambda'(x).$$

Here $g'(x) = 0$ whenever $\lambda'(x) = 0$, i.e., g' has zeros in every neighborhood of ∞, and consequently we are not entitled to apply L'Hôpital's rule at all. However, this conclusion seems rather pedantic; let us go ahead anyway. If we cancel the factor $\lambda'(x)$, we obtain

$$\frac{f'(x)}{g'(x)} = \frac{\lambda'(x)}{\lambda'(x)\varphi(\lambda(x)) + f(x)\varphi'(\lambda(x))}.$$

Now $\lambda'(x)$ is bounded (by hypothesis), $\lambda'(x)\varphi(\lambda(x))$ is bounded, $\varphi'(\lambda(x))$ is bounded away from 0, but $f(x) \to \infty$, so $f'(x)/g'(x) \to 0$. Yet $f(x)/g(x) = 1/\varphi(\lambda(x))$ does not approach zero, since $\varphi(\lambda(x))$ is bounded!

3. DISCUSSION

What went wrong? If you will study any proof of L'Hôpital's rule, you will find a place where it used (or should have used) the assumption that $g'(x)$ did not change sign infinitely often in a neighborhood of ∞. Our example shows that, at least sometimes, L'Hôpital's rule actually fails when this hypothesis is not satisfied.

The phenomenon just described was discovered more than a century ago by O. Stolz [1], [2]. His example was $\lambda(x) = \sin x$, $\varphi(x) = e^x$; it has been repeated in all the modern discussions that I have seen. It was wondering whether there *are* any other examples that led to this note.

One can verify that it is the changes of sign of $\lambda'(x)$ that cause the trouble, not the mere presence of zeros of λ'. In other words, if $\lambda'(x) \geq 0$, the cancellation process still leads to a correct result, as Stolz pointed out. However, it seems wildly improbable that an example of either kind will occur in practice, especially for limits at a finite point. Differentiable functions with infinitely many changes of sign in a finite interval are rarely encountered outside notes like this one; all the less, functions with infinitely many double zeros.

4. HISTORY

Guillaume François Antoine de Lhospital, Marquis de Sainte-Mesme (1651–1704) published (anonymously) in 1691 the world's first textbook on calculus, based on John Bernoulli's lecture notes. He seems to have written his name as above, but it is more familiar as L'Hospital (old French spelling) or L'Hôpital (modern French); I prefer the latter, since it stops students from pronouncing the *s* (which Larousse's dictionary says is not to be pronounced).

References

[1] O. Stolz, Ueber die Grenzwerthe der Quotienten, Math. Ann., 15 (1879) 556–559.

[2] —, Grundzüge der Differential- und Integralrechnung, vol. 1, Teubner, Leipzig, 1893, pp. 72–84.

INDETERMINATE FORMS REVISITED*†

1. INTRODUCTION

You must all have seen at least one calculus textbook. It may surprise some of you that three centuries ago no such book existed: the very first book that was in any sense a calculus text was published, anonymously, in 1696, under the rather forbidding title *Analysis of the Infinitely Small* [5]. It was well known in European mathematical circles that the author was a French marquis, Guillaume de L'Hôpital. (I give him the modern French spelling, which at least keeps students from pronouncing the silent *s* in L'Hospital.) The book was hardly easy reading. It began with propositions like this: "One can substitute, one for the other, two quantities which differ only by an infinitely small quantity; or (what amounts to the same thing) a quantity that is increased or decreased only by another quantity infinitely less than it, can be considered as remaining the same." This sort of presentation gave calculus a reputation, which has survived to modern times, of being unintelligible.

Sylvester, writing in about 1880 [10, vol. 2, pp. 716–17], said that when he was young (around 1830) "a boy of sixteen or seventeen who knew his infinitesimal calculus would have been almost pointed out in the streets as a prodigy like Dante, who had seen hell." (Here and now, we would very likely find students of the same age feeling much the same; but Sylvester, in 1870, was teaching students

Math. Mag. **63** (1990), 155–159.

†This article is the text of an invited address to a joint session of the American Mathematical Society and the Mathematical Association of America, January 14, 1989.

who dealt casually with topics that we would now describe as advanced calculus.) When I was in high school (somewhat later), calculus was thought of, by otherwise well-educated people, as being as deep and mysterious as (say) general relativity is thought of today. My parents knew somebody who was reputed to know calculus, but they had no idea what that was (and they were college teachers—of English). Nowadays there are perhaps too many calculus books, but some of the answers that students give to examination questions make me wonder whether the subject has even now become sufficiently intelligible.

In his own time, and for long afterwards, L'Hôpital had an impressive reputation. Today he is remembered only for "L'Hôpital's rule," which evaluates limits like

$$\lim_{x \to 1} \frac{(2x - x^4)^{1/2} - x^{1/3}}{1 - x^{3/4}}$$

(L'Hôpital's own example) by replacing the numerator and denominator by their derivatives and hoping for the best.

L'Hôpital's rule seems to have fallen somewhat out of favor; I have heard it claimed that all it is useful for is as an exercise in differentiation.

It has been known for some time that many of L'Hôpital's results, including the rule, were purchased (quite literally) from John (= Jean = Johann) Bernoulli. Immediately after L'Hôpital's death in 1704, Bernoulli published an article claiming that he had communicated the rule for 0/0 to L'Hôpital, along with other material, before L'Hôpital had published it. This claim was disbelieved for some two hundred years; sceptics wondered why Bernoulli had not advanced his claim earlier. The reason for the delay eventually became clear when Bernoulli's correspondence with L'Hôpital came to light in the early 1900s. Bernoulli gave the rule to L'Hôpital only after L'Hôpital had promised to pay for it, had repeatedly asked for it, and had finally come across with the first installment. We now also know that there are records of Bernoulli's having lectured on the rule before L'Hôpital's book was published.

In the preface to his book, L'Hôpital says, "I acknowledge my debt to the insights of MM Bernoulli, above all to those of the younger [John], now Professor at Groningen. I have unceremoniously made use of their discoveries and of those of M Leibnis [sic]. Consequently I invite them to claim whatever they wish, and will be satisfied with whatever they may leave me." Considering what we now know, this seems somewhat disingenuous, especially since L'Hôpital was clearly unable to discover for himself how to prove the rule of which, as Plancherel once said of his own theorem, he had "the honor of bearing the name."

You can find the whole story in the 1955 volume of Bernoulli's correspondence [7], or in Truesdell's review of the volume [11].

I used to wonder, from time to time, what kind of proof L'Hôpital had used, but never when I was where I could find a copy of his book. Recently I happened to mention this question to Professor G. L. Alexanderson—who promptly produced his own copy. Professor Underwood Dudley, who is more resourceful than I am, also found a copy, and has translated it into modern terminology [4], but retaining its geometric character (L'Hôpital thought of functions as curves). L'Hôpital actually considered only $\lim_{x\to a} f(x)/g(x)$, where a is finite, $f(a) = g(a) = 0$, and both $f'(a)$ and $g'(a)$ exist, are finite, and not zero. In analytical language, what L'Hôpital did amounts to writing

$$\lim_{x\to a}\frac{f(x)}{g(x)} = \lim_{x\to a}\frac{f(x)-f(a)}{g(x)-g(a)} = \lim_{x\to a}\frac{f'(a)+\epsilon(x)}{g'(a)+\delta(x)}(\epsilon \to 0, \delta \to 0) = \frac{f'(a)}{g'(a)}.$$

It is not trivial to extend such a proof to the cases when $f'(a)$ and $g'(a)$ do not exist (but have limits as $x \to a$), or are both zero, or $f(a) = g(a) = \infty$, or $a = \infty$. I do not know when or by whom these refinements were added, but the complete theory was in place by 1880 [8, 9].

2. A COMMON MODERN PROOF

If you saw a proof of L'Hôpital's rule in a modern calculus class, the probability is about 90% that it is Cauchy's proof. This proof appeals to mathematicians because it is elegant, but often fails to appeal to students because it is subtle. It depends on knowing Cauchy's refinement of the mean-value theorem, namely that (with appropriate hypotheses)

$$\frac{f(x)-f(a)}{g(x)-g(a)} = \frac{f'(c)}{g'(c)},\ c \text{ between } a \text{ and } b. \tag{1}$$

Given this, L'Hôpital's rule becomes obvious.

In spite of its elegance, Cauchy's proof seems to me to be inappropriate for an elementary class. Any proof that begins with a lemma like Cauchy's mean value theorem, that says "Let us consider...," repels most students. Students are also uncomfortable with the nebulous point c. They want to know where it is, and feel that the instructor is deliberately keeping them in the dark. Of course, the exact location of c is completely irrelevant (although numerous papers have been written about it).

3. A CAUTION

Cauchy's proof tacitly assumes that there is a (one-sided) neighborhood of the point a in which $g'(x) \neq 0$. Strictly speaking, if there is no such neighborhood, the limit in (1) is not defined, and we would have no business talking about it. However, if f' and g' are given by explicit formulas, they may happen to share a common factor that is zero at a, and the temptation to cancel this factor is irresistible. One can obtain a spurious result in this way [8, 9; 3, p. 124, ex. 24; 1].

Let me give you a specific example, just to emphasize that there is a reason for the requirement that $g'(x) \neq 0$. Let $f(x) = 2x + \sin 2x$, $g(x) = x \sin x + \cos x$; $a = +\infty$. Then $f'(x) = 4\cos^2 x$, $g'(x) = x \cos x$, and $f'(x)/g'(x) \to 0$, whereas $f(x)/g(x)$ does not approach a limit. The trouble comes from cancelling a factor that changes sign in every neighborhood of the point a; it would have been legitimate to cancel a quadratic factor.

Some writers think that the difficulty arises only in artificial cases that would never occur in practice. But then, what happens to our claim to be giving correct proofs?

You might not guess from Cauchy's proof that there is a discrete analog of L'Hôpital's rule; see, for example, [6]. This was known to Stolz in the 1890s, and has often been rediscovered.

4. A MORE SATISFACTORY PROOF

I want now to show you a proof of L'Hôpital's rule that avoids the difficulties of Cauchy's and establishes a good deal more. It may seem more complicated, but not if you include a proof of Cauchy's mean value theorem as part of Cauchy's proof. This proof is also quite old; Stolz knew it, but preferred Cauchy's proof, perhaps because of Cauchy's reputation. It has been published several times by people (including me) who failed to search the literature.

Let us suppose that $f(x)$ and $g(x)$ approach 0 as $x \to a$ from the left, where a might be $+\infty$; it does no harm to define (if necessary) $f(a) = g(a) = 0$. We may suppose that $g'(x) > 0$ (otherwise consider $-g(x)$). Now let $f'(x)/g'(x) \to L$, where $0 < L < \infty$. Then, given $\epsilon > 0$, we have, if x is sufficiently near a, and a is finite,

$$L - \epsilon \leq \frac{f'(x)}{g'(x)} \leq L + \epsilon,$$

$$(L - \epsilon)g'(x) \leq f'(x) \leq (L + \epsilon)g'(x) \qquad (\text{since } g'(x) > 0).$$

Integrate on (x, a) to get

$$-(L - \epsilon)g(x) \leq -f(x) \leq -(L + \epsilon)g(x)$$

(notice that since g increases to 0, we have $g(x) < 0$). Since $-g$ and $-f$ are positive near a,

$$L - \epsilon \leq \frac{f(x)}{g(x)} \leq L + \epsilon,$$

$$\lim_{x \to a} \frac{f(x)}{g(x)} = L.$$

Only formal changes are needed if $a = +\infty$ or if $L = 0$ or ∞.

For the ∞/∞ case, we get, in the same way, with $\delta > 0$,

$$L - \epsilon < \frac{f(a-\delta) - f(x)}{g(a-\delta) - g(x)} < L + \epsilon,$$

$$L - \epsilon < \frac{\dfrac{f(a-\delta)}{f(x)} - 1}{\dfrac{g(a-\delta)}{g(x)} - 1} \cdot \frac{f(x)}{g(x)} < L + \epsilon.$$

Letting $x \to a$, we obtain

$$L - \epsilon \leq \liminf_{x \to a} \frac{f(x)}{g(x)} \leq \limsup_{x \to a} \frac{f(x)}{g(x)} \leq L + \epsilon.$$

Letting $\epsilon \to 0$, we obtain $f(x)/g(x) \to L$.

If it happens that $f'(a) = g'(a) = 0$, one repeats the procedure with f'/g', and so on. If $f^{(n)}(a) = g^{(n)}(a) = 0$ for every n (which can happen with f and g not identically zero), the procedure fails. Otherwise, the limit can be handled more simply in a single step, as we shall see below.

5. GENERALIZATIONS

If f and g are defined only on the positive integers, we can reason in a similar way with differences instead of derivatives to conclude that if the differences of g are positive, and $f(n)$ and $g(n)$ approach zero as $n \to \infty$, then if

$$\frac{f(n) - f(n-1)}{g(n) - g(n-1)} \to L \qquad \text{as } n \to \infty,$$

it follows that $f(n)/g(n) \to L$. This is sometimes called Cesàro's rule. For more dctail, and illustrations, see [6]. Λ possibly more familiar version is as follows:

If $a_n \to 0$ and $b_n \to 0$, and $a_n/b_n \to L$, then

$$\frac{\sum_{k=1}^{n} a_k}{\sum_{k=1}^{n} b_k} \to L.$$

The key point in the proof of L'Hôpital's rule is the principle that the integral of a nonnegative function ($\not\equiv 0$) is positive. More precisely, if $f(x) \geq 0$ on $p \leq x \leq q$

then

$$\int_p^q f(t)\,dt > 0 \qquad \text{if } p < x < q \text{ and } f(t) \not\equiv 0.$$

Repeated integration with the same lower limit has the same property, as we see by rewriting the n-fold iterated integral as a single integral:

$$\frac{1}{(n-1)!}\int_p^x (t-p)^{n-1} f(t)\,dt.$$

This suggests the appropriate treatment of the case of L'Hôpital's rule when $f'(a) = g'(a) = 0$, or more generally when $f^{(k)}(a) = g^{(k)}(a) = 0$, $k = 1, 2, \ldots, n-1$, but not both of $f^{(n)}(a)$ and $g^{(n)}(a)$ are 0. The positivity of iterated integration then yields the conclusion of L'Hôpital's rule in a single step.

An operator that carries positive functions to positive functions is conventionally called a positive operator. If F is an invertible operator whose inverse is positive, we can conclude that if $F[f(x)]/F[g(x)] \to L$ and $F[g] > 0$, then $f(x)/g(x) \to L$.

As an example of the use of operators, consider $D + P(x)I$, where $D = d/dx$ and I is the identity operator. This is the operator that occurs in the theory of the linear first-order differential equation $y' + P(x)y = Q(x)$. The solution of this differential equation, with $y(a) = 0$, is

$$y = \exp\left\{\int_a^x P(t)\,dt\right\}\int_a^x Q(t)\exp\left\{\int_a^t P(u)\,du\right\}dt. \tag{2}$$

In other words, (2) provides the inverse Λ of $D + P(x)I$.

The explicit formula shows that if $Q(x) \geq 0$ we have $\Lambda[Q] > 0$, so that Λ is a positive operator. Hence we may conclude that if

$$\frac{[D + PI]f}{[D + PI]g} \to L$$

and $[D + PI]g > 0$ then

$$(L - \epsilon)[D + PI]g < [D + PI]f < (L + \epsilon)[D + PI]g.$$

A positive linear operator evidently preserves inequalities. Consequently, if we apply Λ to both sides, we obtain

$$(L - \epsilon)g(x) < f(x) < (L + \epsilon)g(x)$$

and hence

$$f(x)/g(x) \to L.$$

Thus $D + P(x)I$ can play the same role as D in L'Hôpital's rule. It is at least possible that $D + P(x)I$ might be simpler than D.

Since some forms of fractional integrals and derivatives are defined by positive operators, one could also formulate a fractional L'Hôpital's rule.

References

[1] R. P. Boas, Counterexamples to L'Hôpital's rule, *Amer. Math. Monthly* 94 (1986), 644–645.

[2] —, *Indeterminate Forms Revisited*, videotape, Amer. Math. Soc., Providence, RI, 1989.

[3] R. C. Buck and E. F. Buck, *Advanced calculus*, 3rd ed., McGraw-Hill, New York, etc., 1978.

[4] U. Dudley, Review of Calculus with Analytic Geometry, *Amer. Math. Monthly* 95 (1988), 888–892.

[5] [G. de L'Hôpital], *Analyse des infiniment petits, pour l'intelligence des lignes courbes*, Paris, 1696; later editions under the name of le Marquis de L'Hôpital.

[6] Xun-Cheng Hwang, A discrete L'Hôpital's rule, *College Math. J.* 19 (1988), 321–329.

[7] [O. Spiess, ed.], *Die Briefwechsel von Johann Bernoulli*, vol. 1, Birkhäuser, Basel, 1955.

[8] O. Stolz, Über die Grenzwerthe der Quotienten, *Math. Ann.* 15 (1889), 556–559.

[9] —, *Grundzüge der Differential- und Integral-rechnung*, vol. 1, Teubner, Leipzig, 1893.

[10] J. J. Sylvester, *The Collected Mathematical Papers*, Cambridge University Press, 1908.

[11] C. Truesdell, The new Bernoulli edition, *Isis* 49 (1958), 54–62.

SECTION 7

RECOLLECTIONS AND VERSE III

Others have written about the history of Mathematical Reviews (referred to as "MR" in the anecdotes), but Boas here gives some of his experiences as Editor during the early years of this publication that is essential for every mathematician.

Before I arrived at MR, Neugebauer had occasion to write to van der Waerden, and of course wrote in English. (It was one of his principles that when he changed countries, he changed his language accordingly.) van der Waerden complained that Neugebauer didn't use his "Muttersprache." (This seemed ironical, since van der Waerden was Dutch.) Neugebauer replied that what mattered was not his mother's language, but his secretary's.

Once, while I was at MR, we received a Chinese journal that was completely in Chinese. I thought that I ought at least to find the authors' names and the titles of the articles, so I took it over to the Applied Mathematics building to get help from C. C. Lin. After he had given me the information I needed, I asked him what he would do if an author's name was an unfamiliar character. His reply was, "I'm supposed to know them all." Apparently there are only a rather small number of characters that can be people's names.

Since classified research was being done in the Applied Mathematics building, visitors had to sign in at the desk and be escorted to the office they wanted to visit. I thought this system was rather silly, so one day I signed in as V. M. Molotov, in cyrillic letters; nobody noticed.

In the early days of MR, potential reviewers were sent a postcard that contained, among other things, the line "I can review papers in the following languages." Several people took this to mean that they were able to write their reviews in the languages they listed. In fact, in the early days MR accepted reviews in English, French, German, or Italian (the languages of the International Congress of Mathematicians), and a few reviewers made it a point to write the review in the language of the paper, whereas it seemed to MR that it might be more useful to have the review in a language different from that of the paper.

Sometime around 1948, Mark Kac visited me in Cambridge and said, "Ralph, have you noticed how large Mathematical Reviews is getting?" He hadn't stopped to think that I was the person who was having to cope with the growth.

C. R. Adams, who was chairman of the Brown University Mathematics Department, was a stickler for protocol. When I was working for MR, I was living in Cambridge and commuting to Providence by train. Adams would see me coming to work at ten o'clock, and complained to the AMS. Of course, Neugebauer didn't care when I came to work, as long as the work got done; in fact, he told me that

I didn't need to come in every day if there wasn't anything to do; but that never happened.

A friend of mine wanted to find out what was in a paper in Danish. He knew no Danish but tried to decipher the paper on the principle that short words can be neglected. It was hard work, because he neglected the word "ikke," which means "not."

While I was working at MR, I got a letter from the editors of the *Encyclopædia Britannica* asking for advice. The *Encyclopædia* had stated that $\pi/\sin(\pi - z) = \Gamma(z)\Gamma(1 - z)$ and someone had sent the editors numerical results to show that the formula was incorrect; so what about it? It seemed clear that somebody had written $\pi/\sin(\pi \cdot z)$ with a smudged dot that the printer had read as a minus sign.

Although MR's stated policy was that reviews should be descriptive but not critical, a few vicious comments have crept in from time to time. One review read as follows: "This paper contains two theorems. The first is due to the referee and the second is wrong." Another one: "The statements of the results occupy only slightly less space than their proofs, and the reviewer feels that both might have been omitted with advantage."

MR sent me a paper by a Japanese author who kept referring to "stricken mass distributions." I couldn't figure out what those were, and finally wrote to the editor of the journal in which the paper had appeared. He sent me a copy of the referee's report, which had been sent to the author; this said, in part, "The term 'generalized mass distribution' is no longer used. The word 'generalized' should be stricken."

Many mistakes have resulted from "faux amis," words that look alike in different languages but actually have very different meanings. For many years, mathematical papers were written in English about "faculty series" because the German Fakultät means both "factorial" and "faculty."

I have heard radio announcers introducing "L'enfant prodigue" as "The prodigious infant" and as "The infant prodigy." (It should, of course, be "The prodigal son.")

Maurice Heins once showed me a letter he had received from Plancherel that referred to the theorem "dont j'ai l'honneur de porter le nom." It turns out that this was apparently actually a quotation from a quotation. (Somebody told me it was actually first said by Sturm.)

Inflation

The growth of mathematics is producing discontent:
The inflation per annum is pushing ten percent.
Faced with so much information, it's not easy to succeed
In locating any theorem that you appear to need.
The plethora of indexes can leave you in the lurch,
Since it takes less time to prove it than to undertake a search.
I can offer a solution, but it's totally upsetting:
We need to introduce a way of constantly forgetting
The results that won't be needed for another twenty years,
At the end of which they'll surface with appropriate loud cheers.
While the ones that won't be needed for forever and a day,
Once their authors get their tenure will be firmly thrown away.

Editorial Policy

"Let me make it clear to you
This is what we'll never do."*

It really doesn't matter if you don't know how to spell:
You'll find that many readers understand you just as well;
And once the spelling's gone to pot, why then I rather guess
It doesn't really matter if your syntax is a mess;
  But not in *my* journal.

You really needn't worry that it's all been said before,
Since checking out the sources can be an awful bore.
And don't be very troubled if you seem to plagiarize:
If you copy something really good, you just might win a prize;
  But not in *my* journal.

*W. S. Gilbert, *The Mikado*

We often see that authors, even those whose work is strong,
They sometimes go too far and say a thing or two that's wrong.
You needn't worry very hard about a stray mistake:
If you can fool the referee, what difference does it make?
  But not in *my* journal.

Oh, inspiration's wonderful, but second thoughts are best,
And don't you think we must have made those silly rules in jest?
You seekers for perfection, who've made an utter goof,
You'll have a chance to fix it up: rewrite in galley proof;
  But not in *my* journal.

Dialog

Dear Editor:
  Please publish, with least possible delay,
This paper, so the world can know the things I have to say.

Dear Author:
  Though I grant you that precision's all the rage,
I must deplore so many definitions on a page.
You use a new notation that, as far as I can tell,
Combines the worst of Chess and Dance with bits of APL.
Our readers will not stand for this beyond a page or two;
There's just no point in printing it; I send it back to you.

Dear Editor:
  Definitions? I have only 54.
Notation? Mine is clearer—far—than any used before.
If you take time to master it, you'll be amazed to see
How very, very elegant my theory will be.
I know that I'm the very first who ever got it right;
It's going to be a classic that everyone will cite.
So won't you reconsider? It means a lot to me.

Dear Author:
  I'm not editing to raise your salary.
I won't accept this paper, and let me make this clear:
As long as I'm the editor, I make the rules 'round here.

The Author and the Editors

I send the paper in:
They say it is too thin.
When I've corrected that
They say it is too fat.

The next word that I hear
"It isn't really clear."
But after I explain
They look at it again.

At last they write to say
There's been so much delay
That with regret they find
(And hope I will not mind)—
Admittedly, it's sad—
But someone else just published everything I had!

Colloquium Lecture

(To the tune of "There's a Tavern in the Town")

He says it's much too simple in the book;
He wants to give it all a modern look.
In doing so, of course he has to call
On theories I didn't know at all.

To do it right is really rather tough
Unless you're up on all the latest stuff.
The modern methods are so very deep,
The audience was quickly put to sleep.

He used a lot of newly-minted terms
That came too fast for all us stupid worms;
And all that he accomplished was to add
To troubles that I didn't know I had.

Oh, Math is harder than you'd ever, ever guess,
And now my mind is in an awful mess.
I need to know so much to get it right,
I may as well give up the fight.

So I'm about to transfer to a dead
Old language where there isn't much to dread.
Although it's not exciting, still they claim
The words forever stay the same.

Triolet

A theorem of mine
Became just a conjecture:
It sounded so fine
(A theorem of *mine*),
But it died on the vine.
In the course of a lecture,
A theorem of mine
Became just a conjecture.

Please Feed the Archives

It is saddening to dream on
What was lost when Bernie Riemann
Used his letters from Cauchy to light a fire.

If Weyl had left a note
On what Noether never wrote,
Higher algebra might be a good deal higher.

If you want to stake a claim
To everlasting fame,
Then cherish every single bit of writing,

For your discards won't come back,
And they generate a lack
Of information, that the Future will be fighting.

If you just will act with prudence
Some future thesis students
Can hope to clarify a minor mystery;

And what you would have thrown
Away; when it is known,
Will assure you of a footnote in their history.

Though the letters in your "files"
May be just untidy piles,
They throw light on all your failures and successes;

So please do nothing rash
With your daily office trash,
But ship it to the archives down in Texas.

Never throw the stuff away;
Just send it, come what may,
To the celebrated archives down in Texas!

Manuscript Found in a Page Proof

Wynkyn had a lemma, put it in a letter;
Nodd and Blynkyn took it out and made it even better.
As it grew more general, folks added groups and rings,
Convolutors, hyperspinors, lots of other things,
The co- of this, the pseudo- that, and -orphisms galore,
The most of which had not been even dreamt about before.
The Russians now are holding, not just one, but maybe ten,
Sophisticated seminars on Teoriya ВБН.
The founders do not recognize their theory any more,
But they did get advantages they never had before.
Administrations chop and change, but *they*'ll have all survived:
When you're in a Russian acronym, you really have arrived.

SECTION 8

COMPLEX VARIABLES

YET ANOTHER PROOF OF THE FUNDAMENTAL THEOREM OF ALGEBRA*

The following proof of the fundamental theorem of algebra by contour integration is similar to Ankeny's [1], but is simpler because it uses integration around the unit circle (which is usually the first application of contour integration) instead of integration along the real axis; thus there is no need to discuss the asymptotic behavior of any integrals.

Let $P(z)$ be a nonconstant polynomial; we are to show that $P(z) = 0$ for some z. We may suppose $P(z)$ real for real z. (Indeed, otherwise let $\bar{P}(z)$ be the polynomial whose coefficients are the conjugates of those of $P(z)$ and consider $P(z)\bar{P}(z)$.) Suppose then that $P(z)$ is real for real z and is never 0; we deduce a contradiction. Since $P(z)$ does not either vanish or change sign for real z, we have

$$\int_0^{2\pi} \frac{d\theta}{P(2\cos\theta)} \neq 0. \tag{1}$$

But this integral is equal to the contour integral

$$\frac{1}{i}\int_{|z|=1} \frac{dz}{zP(z+z^{-1})} = \frac{1}{i}\int_{|z|=1} \frac{z^{n-1}\,dz}{Q(z)}, \tag{2}$$

Amer. Math. Monthly **71** (1964), 180.

where $Q(z) = z^n P(z + z^{-1})$ is a polynomial. For $z \neq 0$, $Q(z) \neq 0$; in addition, if a_n is the leading coefficient in $P(z)$, we have $Q(0) = a_n \neq 0$. Since $Q(z)$ is never zero, the integrand in (2) is analytic and hence the integral is zero by Cauchy's theorem, contradicting (1).

Reference

[1] N. C. Ankeny, One more proof of the fundamental theorem of algebra, this *Monthly*, **54** (1947) 464.

WHEN IS A C^∞ FUNCTION ANALYTIC?*

Students often wonder why their teachers insist on proving that the remainder in a Taylor expansion approaches zero before they will accept that the series represents the function from which it was obtained. The students tend not to be enlightened by the usual practice of invoking the Lagrange form of the remainder, with its mysterious unspecified intermediate point. It is not only students who are confused: a quite distinguished mathematician (A. Pringsheim) once made a mistake about remainders. This article is the story of the legacy of that mistake, which remained unnoticed for forty years.

The Taylor series of a C^∞ function f can have two kinds of singular behavior: the series may diverge except at its center, or it may converge, in a neighborhood of its center, to a function that differs from f in arbitrarily small neighborhoods of the center. There are many examples of the first kind of singularity, although they are not easy to show to an elementary class. For the second kind, the standard example is the function F defined by $F(x) = \exp(-1/x^2)$ for $x < 0$; $F(x) = 0$ for $x \geq 0$. (As late as 1928 Osgood [5], p. 125, felt constrained to emphasize the triviality of the objection that F "is not really a function" because it is not defined by a single formula.)

If we write the Taylor series for F about various points x, the radius of convergence $\rho(x)$ of the series approaches 0 as $x \to 0$ from the left. Suppose, however, that a function f has the property that $\rho(x)$ is bounded away from 0: $\rho(x) \geq \delta > 0$ for all x in an interval. Is such a function real-analytic (equal to its Taylor series), or might it have a singularity of the second kind? This question was apparently first

*_Math. Intelligencer_ **11**, no. 4, (1989), 34–37. Reprinted by permission.

asked by Pringsheim [6] in 1893. He presented a proof that such a function f is necessarily analytic; I quote it in somewhat modernized notation. Suppose that

$$\sum_{n=0}^{\infty} f^{(n)}(t)r^n/n!$$

converges for $a \le t \le b$ and $0 \le r \le r_1$; then

$$\lim_{n\to\infty} f^{(n)}(t)r^n/n! = 0 \tag{A}$$

for $a \le t \le b$ and $0 \le r \le r_1$. Hence if $0 < s < r_1$, then

$$\lim_{n\to\infty} f^{(n)}(t+\theta s)s^n/n! = 0 \text{ for } a \le t+\theta s \le b. \tag{B}$$

From (B), Pringsheim concludes that the remainder in the Taylor series of f about t approaches 0, so f is analytic on (a, b).

Did you notice the fallacy?

It is rather surprising that Pringsheim, who was an enthusiast for uniform convergence, failed to notice that nothing tells us that the limit in (A) or (B) is uniform in r (or θ), and hence that his proof was incomplete. To put the point more precisely, to say that $\rho(x) > 0$ is to say that

$$\limsup_{n\to\infty}\{|f^{(n)}(x)|/n!\}^{1/n} < \infty. \tag{1}$$

Pringsheim's hypothesis is

$$\limsup_{n\to\infty}\{|f^{(n)}(x)|/n!\}^{1/n} < M, \tag{2}$$

for some finite M independent of x. It is well known (and follows from the Lagrange form of the remainder in Taylor series) that the condition that $|f^{(n)}(x)/n!|^{1/n}$ is uniformly bounded in a neighborhood of each point of an interval is sufficient for f to be analytic.

Even today, uniform convergence is sometimes felt to be a difficult concept; fifty years ago it was considered even more difficult. In 1932 J. J. Gergen had to give a course on Fourier series without using uniform convergence, because the Harvard mathematics department considered the concept to be too difficult for undergraduates.

I was one of those undergraduates. I had happened to read Pringsheim's paper, and was intrigued by the theorem. At first reading, I accepted Pringsheim's proof. However, during the summer vacation I tried to reconstruct the proof, but couldn't make it work. Eventually I gave up, drove 90 miles to Harvard, and looked up the proof. Then I read it more carefully, and realized that it was fallacious.

What do you do when you find a flaw in the proof of an interesting theorem? You might well write to the author, but that is not something that an undergraduate would be likely to do. In any case, I would probably have assumed that Pringsheim was dead after 50 years (actually he was only 82 in 1932, and he lived until 1941). But certainly you would try to find a correct proof.

In 1932–33 I knew only the amount of set theory that occurs in introductory courses in real and complex analysis. In particular, I had never seen the Baire category theorem, but I managed to discover it for myself (only for the real line; it was at least two years before I was to be introduced to complete metric spaces). By using Baire's theorem, I was able at least to prove that if $\rho(x)$ is strictly positive for every x in an interval J, then f is analytic in J except for a nowhere dense closed subset. This result was new, and is interesting in itself, because it shows that a C^∞ function cannot have singular points of the second kind at every point of an interval, or even at the points of a dense subset.

I realized that such a powerful result as Baire's theorem could hardly be new, but it was a long time before I could find it in a book. When I described the theorem to members of the faculty, they didn't recognize it either; however, it is quite likely that I didn't explain it very well.

It was not until early in 1934 that I was able to find a convincing proof of Pringsheim's theorem. I worked it out in a garden on the island of Madeira, which I was visiting for nonmathematical reasons. Then I learned how difficult it is to convince people of the incorrectness of an alleged proof of a correct theorem. When I published my proof in 1935 [1], I ought to have included an analysis of Pringsheim's proof, but I was too naïve to do so. I seem to have thought that of course all those experienced mathematicians out there would see that it was incorrect as soon as they looked at it. This was a fallacy: the reviewer for the *Jahrbuch* didn't believe that Pringsheim's proof was incorrect, and said so. I was eventually able to convince him, and he published a correction to his review. The reviewer for the *Zentralblatt* was more cautious, but I think he didn't believe me either.

My original proof was unnecessarily long-winded; there is a more condensed proof in [7], and another proof [9] by Z. Zahorski, who independently discovered and corrected Pringsheim's error. Also there is a modern proof of Pringsheim's theorem by M. J. Hoffman and R. Katz in [3]. Recently A. Boghossian and P. D. Johnson [2] rediscovered Pringsheim's theorem. They found numerous interesting generalizations and analogous results, and placed the theorem in a broader setting. Any future work in this area must start from [2]. Here I shall only outline a proof in order to indicate the principles on which all existing proofs depend. The argument actually establishes a more general result.

THEOREM. *Let f be C^∞ on an open interval J and let $\rho(x)$ be the radius of convergence of the Taylor series of f about the point x. Suppose that* (1) $\rho(x) > 0$ *at each point x of J, and that* (2) *for every point p of J we have* $\liminf_{x \to p} \rho(x)/|x-p| > 1$. *Then f is analytic in J.*

This contains Pringsheim's theorem because if $\rho(x) \geq \delta > 0$ then

$$\liminf_{x \to p} \rho(x)/|x-p| = \infty.$$

In informal language, the hypotheses of the theorem say that if there is a point p at which f is not analytic then the interval of convergence of the Taylor series of f about points close to p must not extend beyond p.

The proof falls naturally into several steps.

I. Under hypothesis (1) alone, f is analytic on a dense subset of J. This is a straightforward application of Baire's theorem. The hypothesis (1) means that

$$1/\rho(x) = \limsup_{n \to \infty} |f^{(n)}(x)/n!|^{1/n} < \infty.$$

This amounts to saying that there is a finite function μ such that

$$|f^{(n)}(x)| \leq n![\mu(x)]^n, \qquad (n = 1, 2, \ldots).$$

Let E_m be the subset of J on which $m \leq \mu(x) < m+1$, so that $J = \cup_{m \geq 0} E_m$. Baire's theorem says that some E_m fail to be nowhere dense in J; that is, there are an integer m and an interval $K \subset J$ such that E_m is dense in K. Then

$$|f^{(n)}(x)| \leq n!(m+1)^n, \qquad n = 1, 2, \ldots, \text{ for } x \in E_m. \tag{$*$}$$

For $x \in K \backslash E_m$ the inequality $(*)$ holds by continuity. Hence the inequalities in $(*)$ hold for all $x \in K$. This implies that f is analytic in K. Arguing similarly for intervals in $J \backslash K$, we have f analytic on a dense subset of J. Let H be the relative complement of this open set.

II. H contains no isolated points. This step is not essential, but it brings out the role of the second hypothesis in the theorem.

Suppose that y is an isolated point of H. Then y is a common endpoint of two complementary intervals of H, say J_1 on the left and J_2 on the right. If z_1 is a point of J_1, sufficiently close to y, the Taylor series of f about z_1 converges in an interval that extends into J_2; similarly for $z_2 \in J_2$. We can calculate the Taylor coefficients of f at y from the Taylor series of f about z_1, and this series represents f in an interval to the left of y. The coefficients of the same series (about y) can also be calculated from the Taylor series of f about z_2, and this series represents f in an interval to the right of y. Thus f is represented, near y, by its Taylor series centered at y. In other words, y does not belong to H, contrary to assumption. Consequently, H is a perfect set.

III. Since H is closed, we can regard it as a complete metric space and apply Baire's theorem to it. We then find that there is a closed subset P of H and a finite λ such that

$$|f^{(n)}(x)| \le n!\,\lambda^n, \qquad n = 1, 2, \ldots, \tag{3}$$

for all x in the intersection of P with some interval J_1.

IV. We now have (3) satisfied at the points of a nowhere dense closed set P_1. It is easy to see that f is not only analytic in each complementary interval of P_1, but is in fact represented in these intervals by its Taylor expansion about an endpoint r of the interval. We now need to estimate $|f^{(n)}(x)/n!|^{1/n}$ in the complementary intervals to P_1 in J_1. This can be done by direct computation with series (see [2]). Alternatively, we can use an idea of Salzmann and Zeller [7]: extend f to the complex plane and apply Cauchy's estimates for derivatives. By either method, we can obtain a uniform bound in J_1 for $|f^{(n)}(x)/n!|^{1/n}$, and this shows that f is analytic in an interval that contains points of H. This contradicts the definition of H, so H must be empty. That is, f must be analytic on J. This is the conclusion of Pringsheim's theorem.

Pringsheim's theorem seems not to be widely known; I have never seen it mentioned in a textbook. For half a century, I believed that it was a really "pure" theorem, with no applications, either in mathematics or elsewhere. However, a special case has recently found applications in physics ([3], [8]), although the physicists had to discover it for themselves. The result they needed is that if $\liminf_{n\to\infty} |f^{(2n+1)}(t)|^{-1/n} \ge C > 0$ for all t in an interval J, then f is the restriction of an entire function. This condition is equivalent to $|f^{(2n+1)}(t)|^{1/(2n+1)} \le L$ for all $t \in J$. A theorem of Hadamard's then shows that $f^{(2n)}(t)$ satisfies the corresponding inequality (with a larger L). Rather than deducing this from Hadamard's theorem, I shall simply prove it in the required form. By Taylor's theorem with remainder of order 2, if t and $t + \lambda$ belong to J, we have

$$\begin{aligned} f^{(2n)}(t) = {} & \frac{f^{(2n-1)}(t+\lambda) - f^{(2n-1)}(t)}{\lambda} \\ & - \frac{1}{2} f^{(2n+1)}(t + \theta\lambda)\lambda, \qquad |\theta| < 1, \end{aligned}$$

so that $|f^{(2n)}(t)| \leq 2|\lambda|^{-1}L^{2n-1} + |\lambda|L^{2n+1}/2$. Let h be a positive real number less than half the length of J. We may suppose that $L > 2/h$. Then for $t \in J$ one of $t \pm 2/L \in J$, so we may take $\lambda = \pm 2/L$ to obtain

$$|f^{(2n)}(t)| \leq 2L^{2n},$$
$$|f^{(2n)}(t)|^{1/(2n)} \leq 2^{1/(2n)}L < 2L.$$

I am indebted to Professor Johnson for letting me see the manuscript of [2], and also for helpful comments on an earlier draft of the present paper.

References

[1] R. P. Boas, A theorem on analytic functions of a real variable, *Bull. Amer. Math. Soc.* **41** (1935), 233–236.

[2] A. Boghossian and P. D. Johnson, Jr., Pointwise conditions for analyticity and polynomiality of infinitely differentiable functions, *J. Math. Analysis and Appl.* **140** (1989), 301–309.

[3] M. J. Hoffman and R. Katz, The sequence of derivatives of a C^∞-function, *Amer. Math. Monthly* **90** (1983), 557–560.

[4] P. Kolar and J. Fischer, On the validity and practical applicability of derivative analyticity relations, *J. Math. Phys.* **25** (1984), 2538–2544.

[5] W. F. Osgood, *Lehrbuch der Funktionentheorie*, vol. I, 5th. ed., Leipzig and Berlin: Teubner (1928).

[6] A. Pringsheim, Zur Theorie der Taylor'schen Reihe und der analytischen Funktionen mit beschränktem Existenzbereich, *Math. Ann.* **42** (1893), 153–184.

[7] H. Salzmann and K. Zeller, Singularitäten unendlich oft differenzierbarer Funktionen, *Math. Z.* **62** (1955), 354–367.

[8] I. Vrkoc, Holomorphic extension of a function whose odd derivatives are summable, *Czechoslovak Math. J.* **35**(110) (1985), 59–65.

[9] Z. Zahorski, Sur l'ensemble des points singuliers d'une fonction d'une variable réelle admettant les dérivées de tous les ordres, *Fund. Math.* **34** (1947), 183–245; supplement, *ibid.* **36** (1949), 319–320.

SIMPLIFICATION OF SOME CONTOUR INTEGRATIONS*

(Coauthored with M. L. Boas)

In introductory courses on complex analysis we learn to evaluate interesting real integrals by integrating around suitable contours. Common examples are $\int_0^\infty x^{-1} \sin x \, dx$, for which we integrate $z^{-1}e^{iz}$ around a semicircle (center at 0, diameter along the real axis); or the Fresnel integrals $\int_0^\infty \sin x^2 \, dx$ and $\int_0^\infty \cos x^2 \, dx$, for which we integrate $z^{-1/2}e^{iz}$ around a quarter-circle. In such problems we end up having to show that

$$\lim_{R \to \infty} R^\lambda \int_0^{\pi/2} e^{-R \sin \theta} \, d\theta = 0,$$

where $\lambda = 1/2$ for the Fresnel integrals, and $\lambda = 0$ for integrals of the form $\int_{-\infty}^\infty e^{ix} R(x) \, dx$, where R is a rational function with the degree of the denominator greater than the degree of the numerator. The integral involved in the limit is not elementary, and has to be estimated by some device before we can say that its limit is 0. The naive argument that $R^\lambda e^{-R \sin \theta} \to 0$ for each θ except 0 is inadequate, since it would purport to show equally well that $R \int_0^{\pi/2} e^{-R \sin \theta} \, d\theta \to 0$, which is false. Indeed, if this were true it would follow that $\int e^{iz} \, dz$ along $|z| = R$ from $\theta = 0$ to $\pi/2$ tends to zero as $R \to \infty$. This would lead to $\int_0^\infty e^{ix} \, dx = \int_0^\infty e^{-y} \, dy$, although the first integral does not converge.

The usual device is to apply the inequality $\sin \theta \geq 2\theta/\pi$ in order to replace the integral by one that can be evaluated explicitly. Students often find this step difficult because they do not know the inequality, and one has to digress in order to prove it.

Amer. Math. Monthly **92** (1985), 212–213.

An easier device is to break the integral into $\int_0^{\pi/3} + \int_{\pi/3}^{\pi/2}$. The second integral causes no difficulty; in the first, since $2\cos\theta \geq 1$ for $0 \leq \theta \leq \pi/3$, we have

$$R^\lambda \int_0^{\pi/3} e^{-R\sin\theta}\, d\theta \leq 2R^\lambda \int_0^{\pi/3} e^{-R\sin\theta} \cos\theta\, d\theta.$$

The new integral can be evaluated explicitly and evidently approaches 0 as $R \to \infty$.

Another approach is to give up the traditional use of circular-arc contours. For the Fresnel integrals (the more difficult case), take the contour to be the square with vertices at $0, R, R + iR$, and iR (initially with a small indentation at 0). The integrals along $(R, R + iR)$ and $(iR, R + iR)$ are

$$\int_0^R (R + iy)^{\lambda-1} e^{i(R+iy)}\, dy \quad \text{and} \quad \int_0^R (x + iR)^{\lambda-1} e^{i(x+iR)}\, dx.$$

The sum of their absolute values does not exceed

$$\int_0^R (R^2 + y^2)^{(\lambda-1)/2} e^{-y}\, dy + \int_0^R (x^2 + R^2)^{(\lambda-1)/2} e^{-R}\, dx.$$

Since $R^2 + y^2 \geq R^2$ and $x^2 + R^2 \geq R^2$ (and $\lambda - 1 < 0$), the sum does not exceed

$$R^{\lambda-1} \int_0^R e^{-y}\, dy + e^{-R} R^{\lambda-1} \int_0^R dx = R^{\lambda-1}(1 - e^{-R}) + R^\lambda e^{-R} \to 0.$$

For integrals $e^{ix}R(x)\,dx$, the rectangular contour is used in [1] and [3].

Still another method [2] is to use the triangular contour with vertices iR, 0, and R. This gives a slightly more complicated equation for the part of the integral that approaches 0, but has only one integral instead of two to estimate.

References

[1] L. V. Ahlfors, Complex Analysis, McGraw-Hill, New York, 1953, p. 127; 1969, p. 157.

[2] H. P. Boas and E. Friedman, A simplification in certain contour integrals, this *Monthly*, 84 (1977) 467–468.

[3] I. Stewart and D. Tall, Complex Analysis, Cambridge University Press, 1983, p. 223.

SECTION 9

RECOLLECTIONS AND VERSE IV

Boas comments in these anecdotes about the life of a professor and teacher, his experiences at Duke and his many years at Northwestern University. The statements he quotes from students sound familiar to anyone who has taught for some years or served as a Department Chair.

Duke University has (or had) two campuses, an East Campus and a West Campus. I noticed that if I was at the north end of East Campus and drove west, I ended up at the south end of West Campus; but if I drove west from the south end of East Campus, I arrived at the north end of West Campus. After puzzling about this for a while, I inferred the existence of a bridge; I looked for it, and of course found it. This shows that topology can be useful.

When Paul Erdős visited Duke University, we took him to lunch, and the first thing that appeared on the table was, naturally enough, a plate of cornbread. Paul looked at it suspiciously and said, with a strong Hungarian accent, "What is that?" We explained, "It's corn bread; made from corn." Erdős said, "In Hungary, we feed to horses."

At Northwestern, I once taught a course in mathematics for nonmathematicians, which students took when they wanted only to fulfill a requirement. At the end of the course, students were asked to fill out a questionnaire, which ended with the question, "What most interested you in this course?" One student answered this with "the professor's collection of bowties."

Like many other people, when I look up a paper in a journal, I like to browse through the volume to see what else is in it, and I have picked up interesting information that way. One day a visitor to Northwestern told George Springer and me a result that he and his thesis director had conjectured, but had been unable to prove. George and I both thought immediately that the result couldn't be true, since if it were it would be in every textbook. Eventually we remembered something that one of us had picked up while looking for something else, and used it to construct a counterexample. We wrote to our visitor, not only giving the example but also explaining how we had found it. He was chagrined by the example (he had tried hard to prove the theorem), but even more by our having found it as a result of "wild and indiscriminate reading," which was contrary to his mentor's principles (the mentor was a student of R. L. Moore's).

Simeon Leland, the Dean of the College of Arts and Sciences at Northwestern when I arrived there, and for a long time afterwards, used to tell this story, which I fully believe. A professor of mathematics accosted the dean on the campus to complain of having been given only a median salary increase. When he complained, the dean told him, "You are the most median professor I ever saw."

Once, when I was chairman at Northwestern, on a hot summer day, I met one of the instructors walking down the hall barefoot. He asked me whether I minded his meeting his class without shoes on. This was in the 1960's; I said that it didn't matter to me. Then I remembered that, 20 years or so earlier, C. R. Adams had rebuked an instructor for not having his shoes properly shined.

Once, when I was department chair, I received, on the same day, two deputations of students. The first ones complained bitterly about their instructor, whom they claimed not to be able to understand. The second ones started by saying that they thought they were going to do something unprecedented: they wanted to tell me what a wonderful instructor they had: he made everything so clear! Same course, same instructor.

Another student deputation wanted to change their instructor; I asked them what kind of grades they were getting. It turned out that they were all making A's. I apologized for the instructor's deficiencies, but told them they obviously could stand that instructor better than anybody else could, so they had better stay with him.

I don't believe in being sarcastic with students; I can remember only one time that I violated this rule. One of my colleagues dropped dead in her apartment, and somebody had to take over her classes. I took over one myself (a course that I had taught and knew thoroughly). One student in the class complained bitterly (in class) that it wasn't fair to change instructors in the middle of the course; it might spoil his grade and keep him from getting into medical school. I just couldn't take this: the class had been told why they had to have a different instructor. I told that student that he could have the same privilege as the instructor: in other words, Drop dead!

Some years ago, I was teaching an ordinary course in calculus to a class of about 30 students. They couldn't do a problem that started with the data that something was inversely proportional to something else, because they didn't know what "proportional" meant. (The one student who did know had been educated in Greece.) The moral seems to be that people who write textbooks need to know what the students' vocabularies are.

A friend of mine overheard, in the subway, two young men talking in the seat behind him. One was saying, "Where I work, we measure things in thousandths of an inch." The other said, "Gee, that's small. I wonder how many of those there are in an inch." The first one replied, "I dunno—must be millions."

One of my colleagues gave as an examination question, "Prove that the harmonic series diverges." One student puzzled over this for a long time, and finally asked, "Is it all right if I prove that the series *con*verges?"

A store was having a clearance sale, with everything marked down by 25%. I bought some things, the checker added them up on one of those electronic cash registers, and then painstakingly multiplied the result by 0.25, with pencil and paper, oblivious of the calculators displayed on an adjacent counter.

From a lecture: "We showed the result to be best possible, but we didn't go beyond that."

We had a young visitor from southeast Asia who knew a lot of mathematics, but was rather naive in other respects, and found the USA confusing. One evening he was evidently trying to make conversation, and said, "In my country, rich people have elephants. Do your rich men have elephants, too?"

A student couldn't do a calculus problem because he didn't know what the word "horizon" meant; it hadn't occurred to him to use the dictionary that the English Department required him to have.

A trigonometry class had a problem which asked for the distance from a battleship to the shore. One student obtained a distance of a tenth of an inch, about which he apparently had no misgivings.

Two answers by students to examination questions:

a. "Neither of these is the same because it is quite obvious that both are different."

b. "The path of a planet is an eclipse of the sun at one focus." The student who submitted this complained about not getting any credit for his answer, because "the only mistakes were a misspelled word and an incorrect preposition."

A friend who worked for a large engineering firm, back in the pre-computer days, told me that the firm supplied the staff with the most versatile slide rules, the kind that could find powers and roots, solve triangles, and so on. The staff made so many mistakes that the management took the slide rules away and replaced them by specially manufactured slide rules that had just two scales, so that they could multiply and divide, but nothing else. The number of errors greatly decreased.

One mathematician of my acquaintance used deliberately to ask students to find the equation of the tangent to a given curve at a point—which was not on the curve. (If you go through the motions, you get a spurious "answer.")

Murray Peshkin listened to an explanation by a student, along the lines of (a long statement), "so" (another long statement). Murray said, "I understand everything except the "so."

Somebody wrote that he had heard that I had a copy of the first million decimal places of π; would I please send him a Xerox copy of the two or three pages containing them. As a matter of fact, I didn't have what he wanted, but I wrote back and invited him to calculate how thick a stack of pages it would take to print a million digits.

A would-be angle trisector made an appointment (via the department secretary) to tell me about his work. I patiently explained that what he wanted to do was impossible, and offered to give him references. His reply was that he knew that, he had even read and understood the proof, but nevertheless he intended to find a Euclidean construction.

It is interesting that Augustus De Morgan had a very similar experience with a crank, who said "Only prove to me that it is impossible, and I will set about it this very evening." (A Budget of Paradoxes, vol. 2, p. 210.)

Somebody once wrote to me suggesting a simple way of testing an integer for divisibility by 7. He had remarked that since an integer is divisible by 10 if and only if its final digit in base 10 is 0, then to test an integer for divisibility by 7, you just have to write it in base seven and look at the final digit.

I don't remember who introduced me to the following routine, but I have used it with many classes. You say that you are going to see whether the students are mathematicians or physicists. You say, pointing, "Suppose this is a kettle of water, and this is a stove. How do you boil the water?" Eventually some hardy soul says, "Put the kettle on the stove." Then you say "Over here is another kettle of water.

How do you boil the water?" Someone always says, "Put the kettle on the stove." Then you say, "You're a physicist. A mathematician puts it here (pointing to the place where the first kettle started), thus reducing the problem to one that has already been solved."

It can be difficult to get a class to see that there is a difference between "the curve has no slope" and "the curve has zero slope." I used to explain it by the actual example of a bookstore that was having a sale that included "books–one dollar," "books—25 cents" and "books–free." The last kind had zero price; priceless books are the kind kept in the rare book room of the library.

Mark Kac once spent a year in Geneva, and his children went to a Swiss school. I visited the family shortly after they had returned to the U.S.; Kac's son, in high school, had been invited to talk to an Education class about the Swiss educational system. He said that he was going to begin by saying that the most obvious difference from the U.S. system was that it was inconceivable in Switzerland that a high school student would be allowed to address a college audience on any subject whatsoever.

R. M. Winger used to say "Professors know lots of things—if you give them time to look them up."

Do you think you are overworked? Read the following letter from Hermite to Stieltjes, 8 May 1890.

"Theses to read, examinations to correct, and lectures to prepare have so often kept me from enjoying the Easter vacation. I can't tell you the effort I am condemned to expend in trying to comprehend projections in Descriptive Geometry, which I detest, and other things, like the annuity formulas in Arithmetic, etc."

Joe Landin and I were asked to visit and evaluate a certain mathematics department. When we talked to the dean, it was clear that what he really wanted was for us to tell him to fire the chairman. After we had talked to the chairman, other members of the department, and students, it seemed clear to us that the chairman was doing an excellent job, so we told the dean so. He fired the chairman.

Excelsior!

Pete from Popponesset was as brilliant as could be;
He wouldn't work on any petty problems, no, not he!
His prof suggested problems; Peter said they were too soft,
For he would hold the flag of great significance aloft.
He scorned Riemann's conjecture—he claimed it was passé;
He wouldn't waste his time on classic stumbling blocks today.
He's watching and he's waiting for a challenge to arise
Whose answer will provide sufficient glory in his eyes.
His classmates were pedestrian; they got their Ph.D.'s;
They solved some open problems, and promotions came with ease;
They've even won some prizes given by the A.M.S.,
But Peter sneers at all of them and mocks at their "success."
He knows that he'll laugh last, not now, but some fine day
The great Oort cloud of good ideas will orbit one his way.

Pereant

I've proved some theorems, once or twice,
And thought that they were rather nice.
My presentations were rejected
By referees who had detected
Those theorems that I thought my own
In journals I had never known,
And in a strange and knotty tongue.

Oh! For a world still fresh and young,
When fame was won by work alone,
Abstracting journals weren't known,
And (if report can be believed)
No information was retrieved,
Nor academic reputations
Achieved by counting up citations.

(Aelius Donatus (fourth century) is quoted by his student St. Jerome as saying "Pereant qui ante nos nostra dixerunt," freely translated as "Damn the guys who published our stuff first.")

The Assistant Professor Blues

(Written between meetings of the College Committee
on Promotion and Tenure)

When I was young and eager,
And got my Ph.D.,
The head of my department
Gave this advice to me.

"Go publish all your findings,
For if you hold them back
They cannot help to keep you
Upon the tenure track.

"But never publish widely,
For that will do you in.
The Dean will surely say that
You spread yourself too thin.

"Be brilliant with the students
And always be their friend,
Since teaching of poor quality
Will sink you in the end.

"But spend no time on students:
Research is what you need,
Put out in published papers
For deans and such to read.

"In English or in History
You'd best be writing books,
Since publishing mere articles
Gets only dirty looks."

But I—I wouldn't listen
To what he had to say;
With disregard of policy
I went my chosen way.

They fired me today.

Indignation

At Y.X. U., if you've not got
A doctorate, you're paid a lot
(As much as seven thousand) less
Than one who has. For, no success
Seen through the regents' blinkered eyes
(Not though you won a Nobel Prize)
Can overcome the lasting shame
Of no D following your name.
Y.X. is real, and not a jest.
(Its name? Supplied upon request).

An Appreciation of Deans

If you want a committee
To work at high speed,
The wise and the witty
Are not what you need.

Just bring in a dean
And a provost or two;
You'll know what I mean
When you see what they do.

These people are busy:
They want to get through
To return to the dizzy
Routines that they do.

They work hard all day,
Never stopping for fun,
And are gone far away
Once the job is all done.

Computers Are Icumen In

There was a time when teachers
Could just get up and talk.
Slateboards were invented;
You wrote on them with chalk.
At least this slowed the lecture
So the audience could try
To take some notes on everything
Before it passed them by.
But slate was too expensive,
So substitutes arose
That scattered dust on everything,
But mostly on our clothes.
(A kind that's best forgotten—
It died while it was new—
The yellow board with purple chalk
That turned your fingers blue.)
The overhead projector
Seems to help, as well it may,
As long as you can make enough
Transparencies each day.
But I have seen the future,
And don't like what I see:
Computers in the classroom
They've come too fast for me.
There's lots and lots of labor,
As far as I can hear,
Creating complex programs that
Make simple topics clear.
There may be an alternative
We shouldn't overlook:
What's wrong with telling students
They have to read the book?

Contemporary Love Song

My love is a computer with a thousand K of RAM.
It never asks for caviar nor yet prosciutto ham.
Champagne and gourmet cooking are the farthest from its thoughts;
The nourishment I give it is a modest 20 watts.
It's never watching TV when I want to have a chat,
And when I'm through conversing, I can turn it off like *that*.
It never brings me flowers or takes me to a show;
It never seeks affection, or other quid pro quo.
It doesn't kiss me gently or cuddle me in bed,
But it has such lovely software to give me thrills, instead.

The BASIC Blues

(Tune: Frankie and Johnny)

Johnny he made out a flow chart
And typed in a program, but then
All that the terminal printed was
"OUT OF SPACE AT 6010."
It was his program but he done it wrong.

Johnny went over his program;
He thought his routines were so thrifty,
But the terminal printed out only
"ILLEGAL STATEMENT AT 5050."
It was his program but he done it wrong.

Johnny got rash and he transferred
Control right out of a loop.
All the output that Johnny got
Was alphanumeric soup.
It was his program but he done it wrong.

Johnny went back to the console
Convinced that his program would run.
The terminal told him "FOR WITHOUT NEXT"
Before any printout begun.
It was his program but he done it wrong.

Johnny he said "This computer
It makes me feel such a fool,
If I have to do any computing
I'll stay with the old slide rule:
It can't talk back; it never does me wrong."

Hints for Programmers

You always should initialize
Or you may get a big surprise
And likewise run the cost up.

Oh, do not recompute, but store:
The system doesn't mind the chore,
But still it runs the cost up.

Regrettably, as you will find
Computers cannot read your mind,
And so they run the cost up.

Try to avoid the inf'nite loop,
For it will get you in the soup
And greatly run the cost up.

Avoid mistakes in programming:
The system doesn't feel a thing,
But how they run the cost up!

Your program time was badly spent
With error rate of 10%,
Since errors run the cost up.

90%—You'd think it should
Mean something that is pretty good,
And yet it runs the cost up.

Some day when I am old and worn
Supercomputers will be born
That take a program one-tenth wrong
And modify it all along
Until it runs as you would hope—
Although you may not care to cope
With how it runs the cost up.

SECTION 10

INVERSE FUNCTIONS

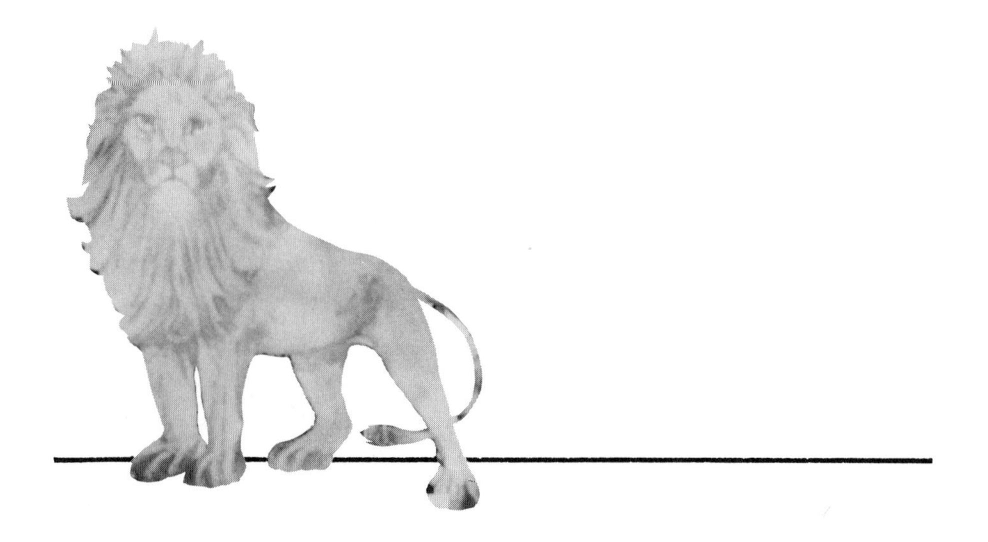

INVERSE FUNCTIONS*

I still remember—after more than half a century—how puzzled I was at first by inverse functions, especially by the notation and the method of getting the graph. If anybody ever told me why the graph of $y = x^{1/2}$ is the reflection of the graph of $y = x^2$ in a 45° line, it didn't sink in. To this day there are textbooks that expect students to think that it is so obvious as to need no explanation. This is a pity, if only because it is such a common practice to define the natural logarithm first and then define the exponential function as its inverse.

This essay is addressed both to students and to teachers. Besides describing how I like to explain inverse functions, I shall exhibit a few of their less well known applications.

1. DEFINITION, TERMINOLOGY, NOTATION

Let us draw the graph of $y = f(x)$, where f is a strictly increasing or strictly decreasing function. Then, as in reading any graph, we start at $(x, 0)$, go vertically to the graph, and then go horizontally to the y-axis to read off y (Figure 1).

If we are given y and want to find x, we reverse the process (Figure 2). From each y we get an x, and this process defines a new function, called the inverse of the function f with which we started. What shall we use as a notation for it? Some standard functions, powers and logarithms, for example, have inverses with their own pet names, but for a general f it is now rather standard practice to call the inverse function f^{-1}. Some books (and now, calculators) avoid the notation $\sin^{-1} x$ by using $\arcsin x$. (In case you wonder what "arc" is doing there, it harks back

---

College Math. J. **16** (1985), 42–47.

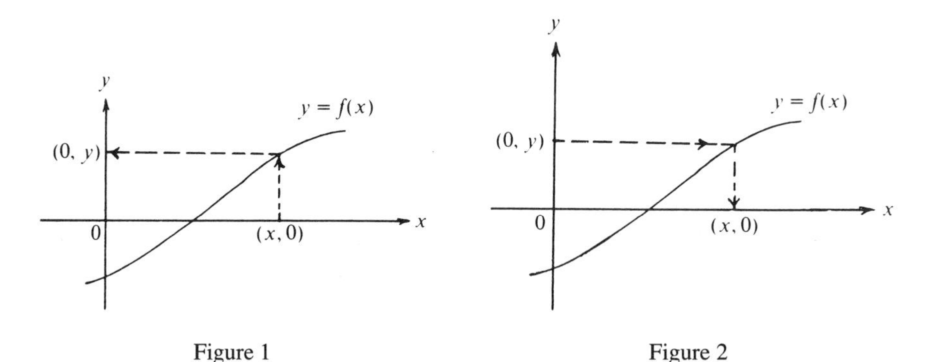

Figure 1

Figure 2

to a convention under which θ stands for the arc of a unit circle subtended by a central angle θ.) There is an additional complication with inverse sines, because the sine function is not monotonic and our construction does not produce an inverse *function* (Figure 3): more than one x can correspond to each y. The best we can do is to pick out a convenient piece of the graph and call its inverse "the principal value of the inverse." In the case of $y = \sin x$, we usually concentrate on the part of the graph where $-\pi/2 \leq x \leq \pi/2$ and call its inverse the principal inverse sine (Figure 4). Logically, it ought to be called the inverse of the principal sine, but it is now too late to try to make the terminology of calculus completely logical. We should, as most modern textbooks do, avoid any idea that a nonmonotonic function has a "multiple-valued" inverse function. The notion of multiple-valued function is best left for a course in complex analysis.

In Figures 2 and 4, we displayed a graphical representation of the inverse function. In the terminology of analytic geometry, Figure 1 is a graph of $y = f(x)$ and Figure 2 is a graph of $x = f^{-1}(y)$. The curves in these two figures are identical; only the interpretation is different. There is no real reason for always graphing $y = f(x)$; in principle, it doesn't matter what letters we use. However, most of us are probably not too well satisfied with Figure 2 as a graphical presentation of the function f^{-1}, because we are conditioned to expect the graph of a function, say, $v = g(u)$ to be drawn with the u-axis positive to the right and the v-axis positive upward. In Figure 2, which is supposed to represent $x = f^{-1}(y)$, the $u(= y)$-axis is positive

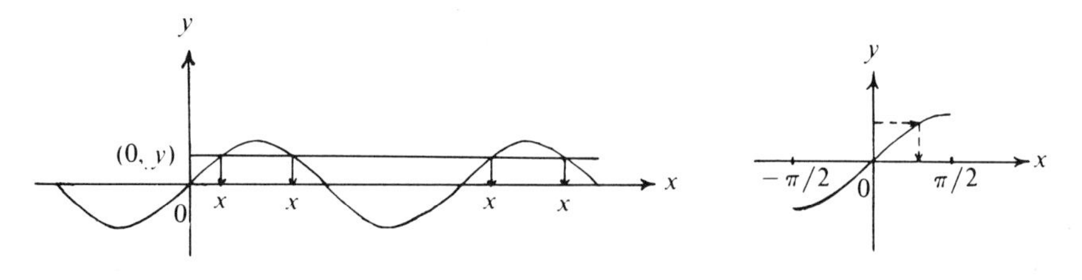

Figure 3

Figure 4

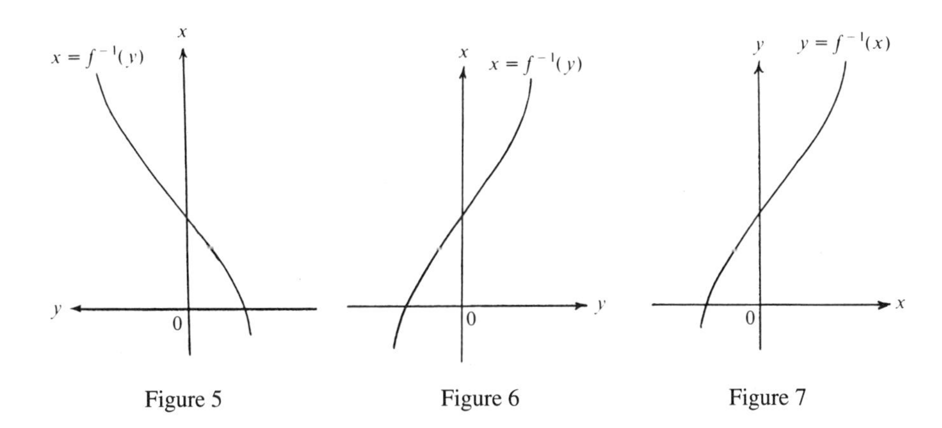

Figure 5 Figure 6 Figure 7

up and the $v(= x)$-axis is positive to the right. Even looking at the graph sideways doesn't help much, because then the x-axis is positive upward, but the positive y-axis points to the left (Figure 5). To make things seem more normal, reflect the picture with respect to the vertical axis (as if there were a mirror there), or look at the graph by looking through the paper from the back. We get Figure 6, with the x-axis positive upward and the y-axis positive to the right. Since it doesn't matter what letters we use, we can change the letters from x to y and from y to x (Figure 7). Now if x is (for instance) 3, the y that we read in Figure 7 is the same as the x that we read from $y = 3$ when we started out with Figure 2.

How can we get from Figure 2 to Figure 7 more simply and with less thought? What we did was, in effect, to interchange x and y in the original equation. This has the same effect as reflecting the graph in the line $y = x$.

To see this, suppose that (a, b) is a point on the graph of $y = f(x)$. Where is (b, a)? First, the line joining these two points has slope -1, so it is perpendicular to the line $y = x$; second, the midpoint of the line segment joining (a, b) and (b, a) is $((a + b)/2, (a + b)/2)$, which is on the line $y = x$. So the line $y = x$ is the perpendicular bisector of the line segment joining (a, b) and (b, a), which is just what we mean by saying that one point is the reflection of the other with respect to the line $y = x$. To perform the reflection in practice, rotate the paper 180° (through space) about the line $y = x$ and then trace the curve and axes on the back of the paper, labeling the horizontal axis x and the vertical axis y.

2. SEMI-f^{-1} GRAPH PAPER

People who deal with experimental data often plot them on "semi-log" paper because equations of the form $y = ab^x$ graph as straight lines; since fitting straight lines to data is a well-established technology, it is easy to determine numerical values for

a and b. Why can't we have a corresponding simplification for other common graphs—as, for example, $y = \sin(ax + b)$?

The answer (of course) is that we can. We cannot expect to convert the whole graph of $y = \sin(ax + b)$ into a straight line, but we can convert a useful piece of it—say, the part for which $ax + b$ is between $-\pi/2$ and $\pi/2$. To see why, and what the question has to do with inverse functions, let us examine the construction of semi-log paper. On this paper (Figure 8), the distances on the horizontal axis are normal (uniform scale), but the scale on the vertical axis looks like the scale on a slide rule: the ordinates on this vertical axis represent the values $\log y$.

Now what happens if we plot the graph of $y = ab^x$ on such graph paper? Since $y = ab^x$ is equivalent to $\log y = \log a + (\log b)x$, we obtain a straight line of slope $\log b$ with intercept $\log a$ on the vertical ($\log y$) axis.

If we recall that $y = \log x$ is the inverse of $y = 10^x$, we see that semi-log paper is constructed by scaling the vertical axis as the inverse function ($y = \log x$) of the function ($y = 10^x$) corresponding to the type of graph we want to linearize (in this case, an exponential function $y = ab^x$).

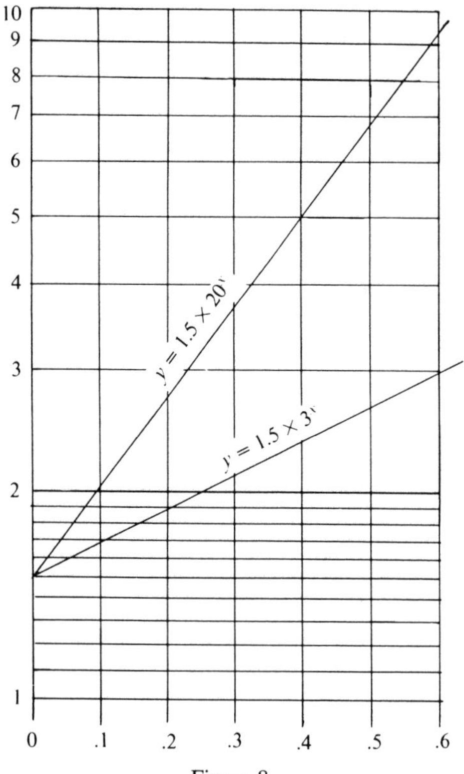

Figure 8

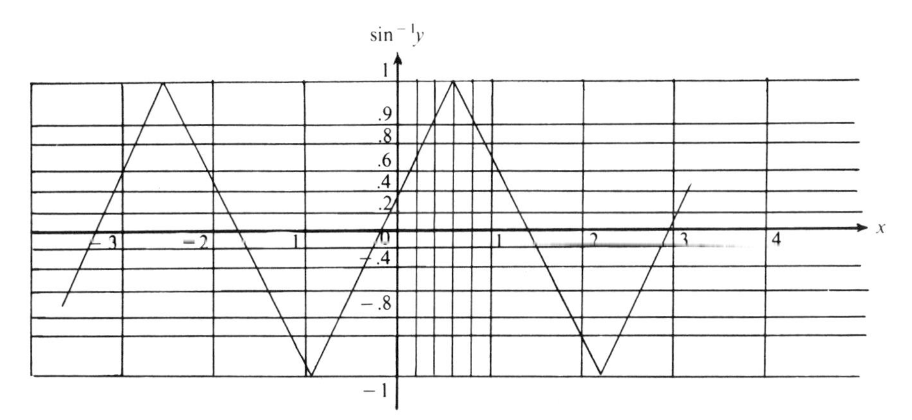

Figure 9. Semi-$\sin^{-1}$ paper showing graph of $y = \sin(2x + 0.4)$.

Nothing prevents our creating, in a similar way, semi-f^{-1} paper for any increasing (or decreasing) function f. For instance, $f(x) = \sin x$ for $-\pi/2 \leq x \leq \pi/2$ has as its inverse function the principal inverse sine, $y = \sin^{-1} x$. Thus, we could print up some graph paper (Figure 9) with the vertical axis marked off as the points $\sin^{-1} y$, where $-1 \leq y \leq 1$. Now if we plot $y = \sin(ax + b)$ for $ax + b$ between $-\pi/2$ and $\pi/2$, the graph on semi-$\sin^{-1}$ paper comes out as a segment of a straight line. This should be clear since $y = \sin(ax + b)$ is equivalent to $\sin^{-1} y = ax + b$. Other parts of the graph will appear as other line segments.

We could do the same thing for $y = x^2$, where $x \geq 0$. The inverse is, of course, $y = x^{1/2}$; so "semi-square-root paper" would linearize the graph of $y = x^2$. Actually one doesn't use semi-square-root paper, but plots $y = x^2$, or more generally $y = ax^n$, on loglog paper, which plots $\log y$ against $\log x$. This is possible only because the logarithm and exponential have nice functional equations, an amenity that we cannot usually expect.

3. INTEGRALS AND AREAS

Sometimes it is worth the mental effort to look at Figures 1 and 2 as graphs of f and f^{-1} without forcing the axes into standard position. The area shaded /// in Figure 10 is $\int_a^b f(x)\,dx - (b - a)f(a)$. If we look at Figure 10 sideways, we see that the area shaded the other way \\\ is $\int_{f(a)}^{f(b)} f^{-1}(y)\,dy - (f(b) - f(a))a$. Since the two areas together make up a rectangle of area $[f(b) - f(a)](b - a)$, we see that

$$\int_a^b f(x)\,dx + \int_{f(a)}^{f(b)} f^{-1}(y)\,dy = bf(b) - af(a).$$

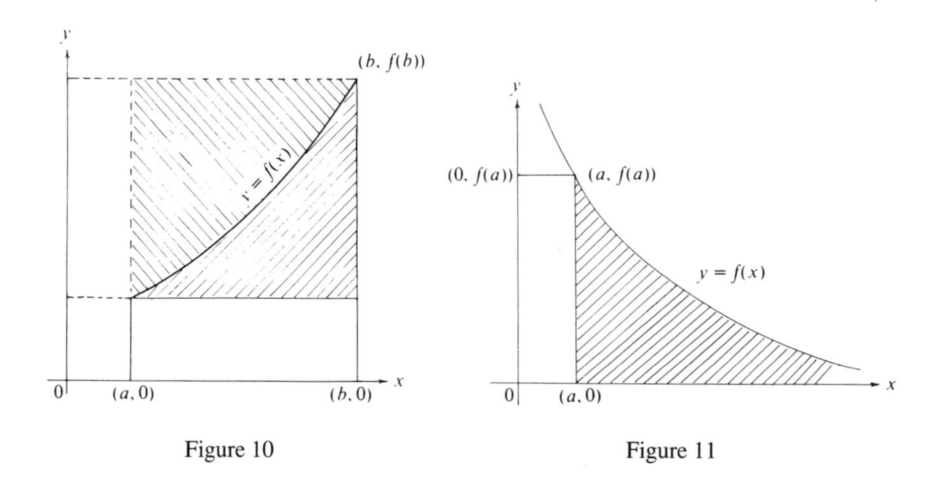

Figure 10

Figure 11

For an illustration of how an unfamiliar integration formula can be deduced from a familiar one by using this formula, see A. D. Holley's paper "Integration by Geometric Insight" [TYCMJ 12, no. 4 (1981) 268–270].

We can do similar things with improper integrals (Figure 11). The shaded area is $\int_a^\infty f(x)\,dx$. We can also think of it as $\int_0^{f(a)} f^{-1}(y)\,dy - af(a)$. Consequently, the first integral (improper at ∞) converges if and only if the second integral (improper at 0) converges. For example, since $\int_1^\infty x^{-p}\,dx$ converges for $p > 1$ and since the inverse of $y = x^{-p}$ is $x = y^{-1/p}$, it follows without further work that $\int_0^1 y^{-1/p}\,dy$ converges for $p > 1$. Equivalently, $\int_0^1 y^{-p}\,dy$ converges for $p < 1$.

4. CALCULATING SUMS

How would you go about calculating the sum $\sum_{k=1}^n [k^{1/2}]$ for a fairly large n, where $[x]$ means the integral part of x? In Figure 12, we have marked the points with two integral coordinates (the so-called lattice points) that are above the x-axis and below or on the curve. The number of these points with abscissa k is $[k^{1/2}]$, so we are asking for the total number of lattice points between 0 and n (excluding 0), and below or on the curve. This looks rather like a crude approximation to the area under the curve $y = x^{1/2}$ from 0 to n, and we might be reminded of Figure 10. With that figure as a guide, let us also mark the lattice points that are above the curve and under the horizontal line $y = n^{1/2}$. Let $m = [n^{1/2}]$. Now we can count the total number of lattice points in Figure 13 in two ways.

First, the number is the total number in the rectangle, namely nm. Second, the number of lattice points is the number on or below the curve, namely (as we just saw) $\sum_{k=1}^n [k^{1/2}]$, plus the number on or above the curve (and inside the rectangle), minus the number on the curve (which we counted twice). We see

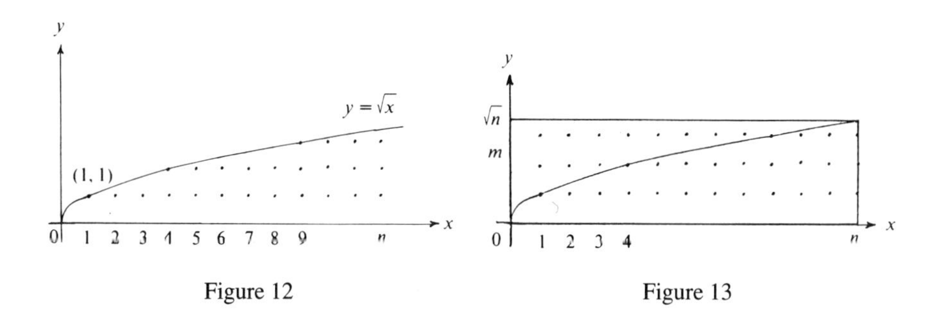

Figure 12 Figure 13

that the number of points above or on the curve is given by $\sum_{j=1}^{m}[j^2] = \sum_{j=1}^{m} j^2$, since j^2 is an integer. It is well known (or verifiable by induction) that $\sum_{j=1}^{m} j^2$ is equal to $m(m+1)(2m+1)/6$. Note also that the number of lattice points *on* the curve is the number of squares that are between 1 and n, inclusive, namely, m. Thus, $\sum_{k=1}^{n}[k^{1/2}] = nm + m - m(m+1)(2m+1)/6$. Since $m = [n^{1/2}]$, this can be written as

$$\sum_{k=1}^{n}[k^{1/2}] = \frac{1}{6}[n](4n - 3[n^{1/2}] + 5),$$

a formula that is quite explicit as long as you accept the bracket function as being explicit. For example,

$$\sum_{k=1}^{100}[k^{1/2}] = \frac{10}{6}(400 - 30 + 5) = 625.$$

This is certainly easier than adding the hundred integers in the sum by hand, although perhaps not easier than programming the addition for a computer.

The same sort of thing can be done with any increasing function f, except that it is unusual to have $f^{-1}(x)$ take integral values at the integers and then be able to produce a simple formula for $\sum f^{-1}(k)$. Try $\sum_{k=1}^{n}[\log_{10} k]$ with $n = 10^6$. (The answer is 5888896.)

INVERSE FUNCTIONS AND INTEGRATION BY PARTS*

(Coauthored with M. Marcus)

Students of calculus often have difficulty with inverse functions; some nontrivial exercises with inverse functions may help. We present here a formulation of integration by parts in terms of inverse functions, which in particular makes some elementary integrals rather easier to evaluate. This is by no means new, although we have not found it in standard calculus books. As an application we give a very short proof of Young's inequality.

Let f be a strictly increasing function with (as is appropriate in calculus) a continuous derivative. Integration by parts tells us that

$$\int_a^b f(x)\,dx = bf(b) - af(a) - \int_a^b xf'(x)\,dx. \tag{1}$$

Let $y = f(x)$, $x = f^{-1}(y)$; then (1) can be written

$$\int_a^b f(x)\,dx = bf(b) - af(a) - \int_{f(a)}^{f(b)} f^{-1}(y)\,dy. \tag{2}$$

Now if $u = f(a)$, $v = f(b)$, (2) becomes

$$\int_{f^{-1}(u)}^{f^{-1}(v)} f(x)\,dx = vf^{-1}(v) - uf^{-1}(u) - \int_u^v f^{-1}(y)\,dy. \tag{3}$$

Thus we can always express $\int f^{-1}(y)\,dy$ in terms of $\int f(x)\,dx$ (or vice versa).

*Amer. Math. Monthly **81** (1974), 760–761.

For example, let $f(x) = \sin x$, $-\pi/2 < x < \pi/2$. Then (3) says that

$$\int_{\sin^{-1} u}^{\sin^{-1} v} \sin x \, dx = v \sin^{-1} v - u \sin^{-1} u - \int_u^v \sin^{-1} y \, dy,$$

whence with $u = 0$,

$$\int_0^v \sin^{-1} y \, dy = v \sin^{-1} v + (1 - v^2)^{1/2} - 1.$$

Those who are familiar with Stieltjes integrals will observe that if we think of (1) as

$$\int_a^b f(x) \, dx = bf(b) - af(a) - \int_a^b x df(x),$$

it continues to hold as long as f is just strictly increasing and continuous; and (2) is still correct in this case (cf. [7], p. 124).

Young's inequality (in the usual form) says that when f is a strictly increasing continuous function with $f(0) = 0$, and $b > 0$, $t > 0$, then

$$bt < \int_0^b f(u) \, du + \int_0^t f^{-1}(y) \, dy. \tag{4}$$

This is geometrically obvious; some recent articles have been devoted to analytic proofs of it (or of its generalizations: the inequality holds—possibly with weak inequality—even when f is weakly monotonic or discontinuous, provided that f^{-1} is suitably interpreted; see [4], [3], [2]). Applications of (4) are given in [5], pp. 111 ff.; [6], p. 49; [1].

The following proof of (4) is very short, but of course depends on our knowing (2).

For $0 < r < b$ it is obvious that

$$(b - r)f(r) < \int_r^b f(u) \, du.$$

Write this as

$$bf(r) - \int_0^b f(u) \, du < rf(r) - \int_0^r f(u) du,$$

and apply (2) to the integral on the right. We get

$$bf(r) - \int_0^b f(u) \, du < \int_0^{f(r)} f^{-1}(y) \, dy.$$

If $0 < t < f(b)$, we can take $r = f^{-1}(t)$, and (4) follows.

Some other applications of (2) are given in [1], and a study of various versions of (4) is given in [2].

References

[1] R. P. Boas and M. B. Marcus, Inequalities involving a function and its inverse, SIAM J. Math. Analysis, 4 (1973) 585–591.

[2] —, and —, Generalizations of Young's inequality, J. Math. Analysis Appl., 46 (1974) 36–40.

[3] F. Cunningham, Jr., and N. Grossman, On Young's inequality, this *Monthly*, 78 (1971) 781–783.

[4] J. B. Diaz and F. T. Metcalf, An analytic proof of Young's inequality, this *Monthly*, 77 (1970) 603–609.

[5] G. H. Hardy, J. E. Littlewood and G. Pólya, Inequalities, Cambridge University Press, 1934.

[6] D. S. Mitrinović, Analytic Inequalities, Springer-Verlag, New York–Heidelberg–Berlin, 1970.

[7] F. Riesz and B. Sz. Nagy, Functional Analysis, Ungar, New York, 1955.

SECTION 11

RECOLLECTIONS AND VERSE V

In this section Boas tells some stories of mathematicians he encountered over many years, both colleagues and those whom he met at meetings and conferences.

The story about Kerékjártó and Bessel-Hagen is well known. If you look in the index of Kerékjártó's book, you will find one citation of Bessel-Hagen; on the indicated page, the name does not appear, but there is a collection of sketches of manifolds, one of which is a pumpkin-head with big ears. There is another joke in the index, much less well known: if you look under "Kerékjártó" there is again one page reference; on that page Kerékjártó's name does not appear, but there is a footnote that says (in translation) "This incorrect theorem was found by me."

J. L. Walsh told this story to illustrate the significance of counterexamples. Reidemeister was writing a paper, and had recently been discussing the subject matter with Study. When he came to one theorem, he couldn't remember whether or not he had discussed that particular one with Study; but he thought, "I'll be nice to the old man," and labeled it "Study's theorem." After the paper was published, somebody constructed a counterexample to the theorem.

There's a theorem in algebraic number theory called "Grünwald's theorem." It had a very long proof, but after a while Whaples found a shorter proof. Later, somebody else found a counterexample.

I knew a mathematician who somehow came to believe that the editors of mathematical journals were conspiring to reject his papers, so he submitted one under a pen name. It was accepted, and then he asked the editors to replace the pseudonym by his real name. The editors had a long argument about whether this was a legitimate request; I believe they finally gave in.

Some years ago, at the end of a summer meeting of the AMS, there was a banquet, at which P. A. Smith was scheduled to give an after dinner speech. When he was called on, he and his wife simply produced recorders and played some music.

There have been many mathematical articles with uninformative titles, but I believe the champion is a paper entitled "On a certain theorem." (It was reviewed in the Jahrbuch über die Fortschritte der Mathematik, but I have lost the reference.)

This story was told to me by someone who claimed to have been present. When Rainich was first in this country, he presented a paper at an AMS meeting.

Afterwards, somebody in the audience got up to say that he thought he had seen something very similar in a paper by Rabinowitz. "Yes," Rainich replied, "*I* am Rabinowitz."

The nineteenth century algebraist H. J. S. Smith is said to have enunciated the principle that a lecture on a piece of research ought to end at the point where the actual research began.

As an example of a completely meaningless coincidence, I offer the fact that $1/(2\pi) = -\log_{10} \ln 2$ with error less than 0.00002.

At an earlier time, I needed to know the value of $\exp(100 - \gamma)$ (γ is Euler's constant), to 60 significant digits, but I had no idea of how to get it. I used to ask people for suggestions, wherever I happened to be, without success. Finally, somebody suggested that I should ask John Wrench. As I subsequently learned, he makes a hobby of high precision computation of interesting numbers. I wrote, and by return got a letter that said "of course" Horace Uhler had computed $\exp(100)$, by hand, long ago; and he (Wrench) possessed a value of $\exp(-\gamma)$ to much more than the necessary precision; so he had only to do one long multiplication, and there was my answer.

At one time, I was involved in some numerical work that called for very high precision multiplication. This was before we had the modern techniques and programs, and I had constructed my own multiplication program. One of the graduate students found me setting up a long multiplication, and said, "Don't you know that there's a FORTRAN program for that? Give me your data, and I'll run it for you." I gave him the data, he ran the program—and got the wrong answer.

Sometimes one has to look at a problem from a less than obvious point of view. Herbert Smith, who was a friend of my family, acquired plans for a sailboat and built it at his summer cottage. Sails are difficult to make, and very expensive to buy, but he discovered that at the time (50 or 60 years ago) Sears Roebuck would make haystack covers to one's specifications, including grommets. So he ordered two triangular haystack covers and sailed the boat for many years with them as sails.

In Hausdorff's *Mengenlehre*, there is a set of axioms that contains a misprint. Somebody wondered what the spaces with the incorrect axiom would be like, and wrote an article about them. Since the "somebody" was R H Bing, the article was quite interesting.

Chandrasekharan told me that what looks like his family name is really his personal name; he said, "Why should I have to go by the same name as all my cousins?"

I once either heard Hermann Weyl say, or read in something he had written, something like "We who cultivate the flower gardens of topology must always remember what a rich and everbearing tree is classical analysis." Later, I wanted to have the reference, but couldn't find the adage in any of Weyl's books, so I wrote, asking him for help. He replied that, although he agreed with the thought, he could not remember having either said or written it, and that if I wanted to quote it, I would have to state it on my own authority. I do so now.

I was once at a party at Hurewicz's apartment, where somebody happened to mention a book on the mathematics of artillery fire. Hurewicz said, "That's a bad book. It ought to be burned." He then hunted until he found the book, went over to the fireplace (where a fire was burning), and proceeded to burn up the book.

There are many mathematical pseudonyms; among them are, of course, the famous "Student"; N. Bourbaki; A. C. Zitronenbaum (who was once listed as a member of the mathematics department at Cornell); John Rainwater; Adam Riese; and O. P. Lossers, who solves problems in journals; "oplossers" means "solvers" in Dutch.

J. D. Tamarkin once told me that he had been asked by a distinguished university to report on a candidate for promotion. He said, "I don't think he's worth a professorship there, but if they want him, who am I to tell them they shouldn't?"

Aurel Wintner came to the United States at the time when newspapers were carrying stories about Einstein, and saying that only a dozen people could understand general relativity. His landlady asked Wintner if he understood it; but when he said

"Yes," she was incredulous: she couldn't believe that she had one of the remarkable dozen staying in her house.

Zeev Nehari told me that he showed a manuscript to Stefan Bergman, and Bergman said "That's a nice paper. Let's make it a joint paper, and I'll write the next one."

Halmos once sent in a review that began "This paper is about valueless measures on pointless spaces." Although this was a perfectly accurate description, I wouldn't let him use it, on the grounds that a reader whose native language was not English might misunderstand it.

Graffiti, Anyone?

Hamilton won acclaim (so people tell)
Carving equations on a bridge; oh, well,
If you or I would try to do that now,
I'd hate to contemplate the awful row.
Surely the cops would come and run us in:
Carving graffiti on a bridge is sin.
Were it allowed, it wouldn't even pay—
Call it a publication? Not today.

Ballade of Old Mathematics

What has become of the rule of three,
  Simple or double, once popular pair?
What is a Napier analogy?
  What was duplation (no great affair)?
Where is the method, hard to instill,
  For finding square roots the clumsy way?
Who learns about latera recta still?
  That was the math of yesterday.

Who learns to find positions at sea
After the method of Saint Hilaire?
Arithmetic classes have long been free
Of cloff and suttle, of tret and tare.
Alligation is over the hill;
Involutes, evolutes, where are they?
Over such losses, whose tears will spill?
That was the math of yesterday.

Slide rules? Dead, as we all agree.
Can you mention surds? I doubt you dare.
Haversines we no longer see.
For finding subtangents, we hardly care.
Horner's method, that used to fill
Many an hour, fills none today.
Triangle solving's an obsolete skill.
That was the math of yesterday.

Envoy

Instructor, ponder this codicil,
An awkward truth that you can't gainsay:
What you're teaching now, with so much good will,
Is tomorrow's math of yesterday.

Getting Even

I
Dorstenia: The Indirect Insult

Suppose another scholar has done some wrong to you;
Don't get a gun and shoot him, for this would never do.
It's easy to destroy him with just a little trying:
You publish wrong conclusions that you thank him for supplying.

Or there's a subtler method over which you ought to mull.
Linnæus had to name a plant he thought was rather dull.
It wasn't ornamental and it hadn't any color:
He named it after Dorsten, who he thought was even duller.

II
Magic Word

Would you like to kill your mother
As many people do?
Our anti-burglar laser
Is just the thing for you.

The kitchen blew your lunch up?
Its maker is to blame.
We've stacks of little letter bombs
So you can do the same.

If your new car's a lemon,
We know a thing or two.
We've vicious propaganda
To make your day for you.

This stuff—we mustn't sell it
To folks like you and you
But call yourself a GOVERNMENT:
We'll put the order through.

Spelling Lesson

Weep for the mathematicians
Posterity acclaims:
Although we know their theorems
We cannot spell their names.

Forget the rules you thought you knew—
Henri Lebesgue has got no Q.

Although it almost rhymes with Birkhoff,
Two H's grace the name of Kirchhoff.

The Schwarz of inequality
And lemma too, he has no T.

The "distribution" Schwartz, you see
Is French, and so he *has* a T.

In Turing's name—no German, he—
An umlaut we should never see.

Hermann Grassmann—please try to
Spell both his names with 2 N's, too.

If you should ever have to quote
A Harvard Peirce, be sure to note
He has the E before the I;
And so does Klein. Rules still apply
To Wiener: I precedes the E;
The same for Riemann, as you see.
But Weierstrass, you must agree,
Has it both ways, with EIE.

Fejér, Turán, Cesàro, Fréchet—
Let's make the accents go that way;
Don't lose the squiggly little bits;
They don't mark stress—they're diacrits.

And as for (Radon)-Nikodým,
Restore the accent, that's my dream.

But there is one I leave to you,
Whatever you may choose to do:
Put letters in or leave them out,
Garnish with accents round about,
Finish the name with -eff or -off:
There *is* no way to spell Чебышёв.

If you would have a friend in me,
Spell "Boas" thus, not with a z.

Naming Things

I

If asteroids were what you find
Then you could give them any kind
Of names that might appeal to you.
Biologists can do so, too.
In mathematics, it's quite tough:
Our options are not broad enough.

II

Why does tradition have to matter?
Why can't we name our finds to flatter
The head of our department, or
An influential senator?
Or give a relative a lift?
(A theorem makes a charming gift.)
"Dear Mary's theorem" sounds quite nice;
"Aunt Emma's lemma" would add spice.

III

Now, if you find a great result,
Why not admit that you exult,
And call it "My ingenious scheme,"
Or label it as "Childhood's Dream"?
Our ancestors were less inhibited,
And let their feelings be exhibited.
Considering what they could do,
It's stuffy editors, that's who,
If we'd attempt to do the same
Would never let us play this game.

SECTION 12

BOURBAKI

BOURBAKI AND ME*

From time to time, I get asked about my "feud" with Bourbaki, of which various versions seem to exist. I should like to set the record straight.

Between 1945 and 1950 I was the Executive Editor of Mathematical Reviews. I suppose that that is why I was asked to write the annual article on Mathematics for the Encyclopædia Britannica Book of the Year. On one occasion, I mentioned as a notable event (because I thought it was) the appearance of several volumes of Bourbaki's magnum opus. The name of Bourbaki was less familiar to the public than it has become subsequently, so I thought it worthwhile to mention that it was a collective pseudonym (as I knew from having met André Weil in 1939, when the group was more open about its activities than it later became).

Presently I received a letter dated "From my ashram in the Himalayas," beginning, "You miserable worm, how dare you say that I do not exist?" and signed by Bourbaki. (Here Bourbaki was displaying less than his usual precision of language, since I had not asserted his nonexistence, only his nonindividuality.) Bourbaki also protested to the Encyclopædia Britannica; the editor (Walter Yust) forwarded his letter to me with a request for clarification. I replied with a brief explanation, adding that if Yust still had any doubts, he might ask Saunders Mac Lane, since both he and Mac Lane were in Chicago, whereas I was in Providence. I had in mind that Yust would probably telephone Mac Lane, but Yust wrote a letter, which Mac Lane showed to Weil, who was then at the University of Chicago. Weil told Mac Lane, apparently quite forcefully, that he *must* tell the Encyclopædia that Bourbaki *did* exist. Mac Lane then wrote a guarded letter that didn't actually say that Bourbaki existed, but hinted strongly that he did. Naturally this also came to me, and I replied

Math. Intelligencer* **8, no. 4 (1986), 84–85. Reprinted by permission.

with more details; and I had the inspiration of referring Yust to J. R. Kline, then the Secretary of the American Mathematical Society. Kline told Yust that Bourbaki had applied for membership in the Society, but had had his application returned with the remark that the American Mathematical Society has two classes of membership: individual and institutional; he added, "I understand that this is not an application from an individual."

That was the last I heard, but I was told later that Bourbaki had tried to float a rumor that Boas was merely a collective pseudonym of the editors of Mathematical Reviews.

It was with some pleasure, years later, that I received an invitation to write a short article about Bourbaki for the Encyclopædia Britannica.

It is worth adding that Bourbaki was the only living individual included in the Dictionary of Scientific Biography.

In the May 1957 issue of Scientific American, Paul R. Halmos wrote an article "Nicolas Bourbaki" in which he explained the lore that had built up about this prolific but nonexistent mathematician. It prompted Boas to write the following letter, which appeared in the "Letters" section of the July 1957 issue of Scientific American.

Sirs:

E. S. Pondiczery of the Royal Institute of Poldavia has asked me to transmit the following communication. Dr. Pondiczery is traveling abroad and does not have your address.

"While I greatly appreciate Paul R. Halmos's flattering references to me in his article on Nicolas Bourbaki [Scientific American, May], I feel bound to admit that Poldavia was Bourbaki's homeland before it was mine; to him goes the credit (if credit it be) for the discovery of this fascinating land.

"May I take this opportunity to point out that on page 94 Halmos perpetrates the howler 'Chevallier' for 'Chevalley.' No real mathematician could do this. Is Halmos a pseudonym for a group of young people who write for *Scientific American*?" Signed: E. S. Pondiczery, Royal Institute of Poldavia (in Exile).

R. P. Boas
Northwestern University
Evanston, Ill.

In 1970 Boas wrote the "official" biography of Bourbaki where he refers to the earlier article in Scientific American by Halmos.

BOURBAKI*

BOURBAKI, NICOLAS. Bourbaki is the collective pseudonym of an influential group of mathematicians, almost all French, who since the late 1930's have been engaged in writing what is intended to be a definitive survey of all of mathematics, or at least of all those parts of the subject which Bourbaki considers worthy of the name. The work appears in installments that are usually from 100 to 300 pages long. The first appeared in 1939 and the thirty-third in 1967; many intervening installments have been extensively revised and reissued. The selection of topics is very different from that in a traditional introduction to mathematics. In Bourbaki's arrangement, mathematics begins with set theory, which is followed, in order, by (abstract) algebra, general topology, functions of a real variable (including ordinary calculus), topological vector spaces, and the general theory of integration. To some extent the order is forced by the logical dependence of each topic on its predecessors. Bourbaki has not yet reached the other parts of mathematics. Although the work as a whole is called *Elements of Mathematics*, no one could read it without at least two years of college mathematics as preparation, and further mathematical study would be an advantage.

The exact composition of the Bourbaki group varies from year to year and has been deliberately kept mysterious. The project was begun by a number of brilliant young mathematicians who had made important contributions to mathematics in their own right. At the beginning they made no particular attempt at secrecy. With the passage of time, however, they seem to have become more and more enamored of their joke, and have often tried to persuade people that there is indeed an individual named N. Bourbaki, who writes the books. Indeed, Bourbaki once applied for membership in the American Mathematical Society, but was rejected on the ground

Dictionary of Scientific Biography, Volume II, Scribner's, 1970, pp. 351–353. Reprinted by permission of the American Council of Learned Societies.

that he was not an individual. The original group included H. Cartan, C. Chevalley, J. Dieudonné, and A. Weil (all of whom are among the most eminent mathematicians of their generation). Many younger French mathematicians have joined the group, which is understood to have ten to twenty members at any one time and has included two or three Americans. The founding members are said to have agreed to retire at the age of fifty, and are believed to have done so, although with some reluctance.

The origin of the name Nicolas Bourbaki is obscure. The use of a collective pseudonym was presumably intended to obviate title pages with long and changing lists of names and to provide a simple way of referring to the project. The family name appears to be that of General Charles-Denis-Sauter Bourbaki (1816–1897), a statue of whom stands in Nancy, where several members of the group once taught. Possibly the Christian name was supposed to suggest St. Nicholas bringing presents to the mathematical world.

In the early days Bourbaki published articles in mathematical journals, as any mathematician would. He soon gave that up, however, and his reputation rests on his books. People who are unsympathetic to the "new mathcmatics" introduced into the schools since 1960 accuse Bourbaki of having inspired that movement. The accusation is probably unjustified, although aspects of his work bear a superficial resemblance to less attractive aspects of new mathematics. Bourbaki himself does not intend his approach to be used even in college teaching. Rather, it is meant to improve a mathematician's understanding of his subject after he has learned the fundamentals and to serve as a guide to research.

The most obvious aspects of Bourbaki's work are his insistence on a strict adherence to the axiomatic approach to mathematics and his use of an individual and (originally) unconventional terminology (much of which has since become widely accepted). The former is the more important. Any mathematical theory starts, in principle, from a set of axioms and deduces consequences from them (although many subjects, such as elementary algebra, are rarely presented to students in this way). In classical axiomatic theories, such as Euclidean geometry or Peano's theory of the integers, one attempts to find a set of axioms that precisely characterize the theory. Such an axiomatization is valuable in showing the logical arrangement of the subject, but the clarification so achieved is confined to the one subject, and often seems like quibbling.

A good deal of the new mathematics consists of introducing such axiomatizations of elementary parts of mathematics at an early stage of the curriculum, in the hope of facilitating understanding. Bourbaki's axiomatization is in a different spirit. His axioms are for parts of mathematics with the widest possible scope, which he calls structures. A mathematical structure consists, in principle, of a set of objects of unspecified nature, and of certain relationships among them. For example, the structure called a group consists of a set of elements such that any two can be combined to give a third. The way in which this is done must be subject to suitable axioms. The structure called an order consists of a set of elements with a relationship between any two of them, corresponding abstractly to the statement (for numbers) that one is greater than the other.

Having studied a structure, one may add axioms to make it more special (finite group or commutative group, for example). One can combine two structures, assuming that the objects considered satisfy the axioms of both (obtaining, for example, the theory of ordered groups). By proceeding in this way, one obtains more and more complicated structures, and often more and more interesting mathematics. Bourbaki, then, organizes mathematics as an arrangement of structures, the more complex growing out of the simpler.

There are great advantages in dealing with mathematics in this way. A theorem, once proved for an abstract structure, is immediately applicable to any realization of the structure, that is, to any mathematical system that satisfies the axioms. Thus, for example, a theorem about abstract groups will yield results (which superficially may look quite different) about groups of numbers, groups of matrices, or groups of permutations. Again, once it is recognized that the theory of measure and the theory of probability are realizations of a common set of axioms, all results in either theory can be reinterpreted in the other. Historically, in fact, these two theories were developed independently of each other for many years before their equivalence was recognized. Bourbaki tries to make each part of mathematics as general as possible in order to obtain the widest possible domain of applicability. His detractors object that he loses contact with the actual content of the subject, so that students who have studied only his approach are likely to know only general theorems without specific instances. Of course, the choice of an axiom system is never arbitrary. Bourbaki's collaborators are well aware of the concrete theories they are generalizing, and select their axioms accordingly.

Bourbaki has been influential for a number of reasons. For one thing, he gave the first systematic account of some topics that previously had been available only in scattered articles. His orderly and very general approach, his insistence on precision of terminology and of argument, his advocacy of the axiomatic method, all had a strong appeal to pure mathematicians, who in any case were proceeding in the same direction. Since mathematicians had to learn Bourbaki's terminology in order to read his work, that terminology has become widely known and has changed much of the vocabulary of research. The effect of the work in the development of mathematics has been fully commensurate with the great effort that has gone into it.

BIBLIOGRAPHY

[I] Original Works. Works by Bourbaki include "The Architecture of Mathematics," in *American Mathematical Monthly*, 57 (1950), 221–232; and *Éléments de mathématique*, many numbers in the series Actualités Scientifiques et Industrielles (Paris, 1939–).

[II] Secondary Literature. André Delachet, "L'école Bourbaki," in *L'analyse mathématique* (Paris, 1949), pp. 113–116; Paul R. Halmos, "Nicolas Bourbaki," in *Scientific American*, 196, no. 5 (May 1957), 88–99.

SONNET

(Lines written after reading too many abstracts of talks at a Mathematics meeting. After Shakespeare, Sonnet 130, "My mistress' eyes are nothing like the sun.")

No diagrams within my work commute;
Language will do. Against the tide of groups,
Lie, semisimple—I'm with King Canute.
Let others prate of posets and of loops,
Functors and morphisms, maximal ideals;
Give me the clichés of an earlier age.
Let no nonstandard models of the reals,
Sur- or bijections decorate my page.
The complex plane contains enough; for me
No sheaves of germs upon a manifold.
I'll never be approved by Bourbaki;
Words grow apace, but still my soul's not sold.
And yet I think my work was as profound
As this, tricked out with terms of modish sound.

(According to the story, King Canute, disgusted by the fulsome flattery of his courtiers, resolved to give them an object lesson. He therefore had his throne set up at the edge of the sea, and forbade the tide to come any further.)

GAME ADJOURNED

Most good chess players start at an early age, but Valentina Pondiczery started at about thirty. True, she was not a complete stranger to the game: her father was an International Grandmaster (Pondiczery's gambit has a minor place in *Modern Chess Openings*), and her mother was at one time women's champion of Poldavia, but Valentina herself was much more interested in algebraic topology. Her achievements in this field were considerable, although too esoteric even for the average professional mathematician to appreciate; they had brought her a Fields Medal, membership in the National Academy of Sciences, and a professorship at Harvard. At that point she felt that she could spare a few years to work in a different field. Partly because she was also seriously interested in Women's Rights, she decided to attack chess, in an endeavor to refute the popular belief that women are not very good at that game. Do not get the wrong idea: Valentina did not propose to do anything as commonplace as becoming a world champion; she intended to do no less than to apply her mathematical talents to producing a strategy for chess. As you know, it is a theorem of game theory that chess, or any similar game, has either a winning strategy for White, or one for Black, or one leading inevitably to a draw; the problem is, which? By a winning strategy for White, I mean that White always has a best move, whatever Black does, and if White always plays that move, White will always win. Furthermore, if White ever fails to play the correct move, Black (knowing the strategy) can then force a win. Since it is unlikely that White will follow the strategy without knowing it, Black (knowing the strategy) is in good shape if White is ignorant, as soon as White makes one bad move.

Valentina gave herself five years; it took her only three. Naturally people have been looking for a strategy in chess for hundreds of years, but no mathematician of Valentina's ability had ever tackled the problem. Unfortunately things are seldom as simple as they appear: with her newfound knowledge, Valentina was very far from instantly becoming the world champion, or even a player of moderately high rating. The strategy was there, all right, but it required the player to make computations with matrices, Lie algebras, and so on, after each move by the opponent. Valentina, after some practice, got it down to where she could usually find the correct move in much less than an hour, but she couldn't do it in her head. Now tournament chess is played under exacting conditions: you have only a limited time per move (say, 40 moves in two hours), and in any case you have to do all your thinking in your head; no auxiliary chess boards, no pencil and paper, no table of Clebsch-Gordan coefficients. The obvious answer is to computerize the calculations, but you can't sit in a tournament with a connection to a computer in your ear. Notice that Valentina had not programmed a computer to play chess, which is quite a different problem; she had merely produced a series of computations that enabled a person using them to play winning chess.

Valentina naturally wanted to try out her strategy in actual play, but this seemed impossible, for reasons that I have just explained. After some thought, she decided that the most promising approach would be to pretend to be a computer program. There are, as you know, a number of chess-playing programs, including computers that do nothing except play chess; they have their own tournaments, and occasionally they are allowed to compete against people in regular tournaments. The usual arrangement is that each program has a human representative who transmits the opponent's move to the computer, receives the computer's move, and moves the piece on the board. Valentina's idea was to insert herself as a link in the communication process; she would sit at the computer, do the necessary computation, and send the result on to her representative at the tournament. Of course it all had to be arranged very secretly. Valentina had to persuade the Computing Center to let her monitor the computer; she explained this unusual procedure on technical grounds that she could get away with only because she knew so much that she could bemuse the Director of the Center with more technical jargon than even he commanded. He was, in fact, happy to do a favor for so distinguished a colleague.

At first all went well. Valentina entered her program in the next computer tournament, and handily outclassed Bell Labs, the previous champion. This was not too surprising, since at the time the best computer program had only an Expert rating, and could be beaten by a player from a good college team. Still, this success gave program BOURBAKI (as Valentina whimsically called it, in honor of the great French mathematician) some standing in chess circles. Her next step was to enter the Greater Boston Open. To do this, she had to persuade the tournament director to accept BOURBAKI as a legitimate player. He was, at first, not impressed by the claims of a mere mathematician, but he gave in after Valentina challenged him to play against the program, and he lost. He was an International Grandmaster, but his days of active competition were long past, although he liked to pretend that he was as good as ever. After he had condescendingly taken the black pieces and seen his Luzhin Defence torn to bits in eighteen moves, he was more than willing to let BOURBAKI try. Indeed, by this time he was secretly hoping that BOURBAKI would demolish some of his hated local opponents. And so it did.

The results of the Greater Boston Open caused a sensation. *Chess Life and Review* ran an article, "Computer chess comes of age." *Shakhmatnyĭ Zhurnal* said, in effect, just wait until Botvinnik's program comes on line. Tournament directors urged BOURBAKI to compete in their tournaments; Valentina did not have time to accept all the invitations. She did enter BOURBAKI in the McDonald's Hamburger Invitational (the tournament with the largest prizes in history), where it neither lost nor drew a single game, against a field of International Grandmasters.

So Valentina had succeeded—or had she? If you have a successful program for something, it's like a proof of a mathematical theorem: you are supposed to make it public. But Valentina didn't have an honest program for playing chess; she had a computerized strategy for playing chess. She could, of course, simply publish the strategy; but that, on reflection, was unthinkable. To publish the strategy would ruin the game. Not completely, of course; the "fish" wouldn't try to understand it, and couldn't understand it if they tried, so they would go on playing as before. But what about the Grandmasters, with their lives invested in the game? All chess players think that they can win every game; sometimes they make mistakes, and lose, but they dream of playing perfectly. Suppose you have the black pieces, White knows the winning strategy, and you know for certain that you will lose if your opponent plays perfectly, whatever you yourself do. Suppose that somebody improves the strategy so that it can be memorized and applied routinely. Did Valentina want to go down in history—chess history, anyway—not as one of the leading topologists of the century, but as the person who destroyed the game of chess? Certainly not.

So BOURBAKI retired from competition. Valentina replied to requests for information by saying that she was working on improvements and would send a copy of the program when it was perfected. Actually she put it in a safe-deposit box (she couldn't bring herself to destroy it). Her will stipulates that it is to remain locked away for a hundred years, after which the president of the Fédération International des Échecs will be allowed to examine it and decide whether the world is ready for it.

Fortunately even the chess world has a short memory. It has forgotten that a computer program once defeated the world's best; there are new prodigies every year. Valentina has taken up Analytic Number Theory, which is even more difficult than chess. She hasn't yet proved the Riemann Hypothesis, but she has come closer than anybody else. Nobody now thinks of her in connection with chess at all, and nobody asks Bourbaki whether he is related to the famous chess player.

SECTION 13

THE TEACHING OF MATHEMATICS

“IF THIS BE TREASON...”*

If I had to name one trait that more than any other is characteristic of professional mathematicians, I should say that it is their willingness, even eagerness, to admit that they are wrong. A sure way to make an impression on the mathematical community is to come forward and declare, “You are doing such-and-such all wrong and you should do it *this* way.” Then everybody says, “Yes, how clever you are,” and adopts your method. This of course is the way progress is made, but it leads to some curious results. Once upon a time square roots of numbers were found by successive approximations because nobody knew of a better way. Then somebody invented a systematic process and everybody learned it in school. More recently it was realized that very few people ever want to extract square roots of numbers, and besides the traditional process is not really very convenient. So now we are told to teach root extraction, if we teach it at all, by successive approximations. Once upon a time people solved systems of linear equations by elimination. Then somebody invented determinants and Cramer’s rule and everybody learned that. Now determinants are regarded as old-fashioned and cumbersome, and it is considered better to solve systems of linear equations by elimination.

We are constantly being told that large parts of the conventional curriculum are both useless and out of date and so might better not be taught. Why teach computation by logarithms when everybody who has to compute uses at least a desk calculator? Why teach the law of tangents when almost nobody ever wants to solve an oblique triangle, and if he does there are more efficient ways? Why teach the conventional theory of equations, and especially why illustrate it with ill-chosen examples that can be handled more efficiently by other methods? As a professional mathematician, I am a sucker for arguments like these. Yet, sometimes I wonder.

Amer. Math. Monthly **64** (1957), 247–249.

There are a few indications that there is a reason for the survival of the traditional curriculum besides the fact that it is traditional. When I was teaching mathematics to future naval officers during the war, I was told that the Navy had found that men who had studied calculus made better line officers than men who had not studied calculus. Nothing is clearer (it was clear even to the Navy) than that a line officer never has the slightest use for calculus. At the most, his duties may require him to look up some numbers in tables and do a little arithmetic with them, or possibly substitute them into formulas. What is the explanation of the paradox?

I think that the answer is supplied by a phenomenon that everybody who teaches mathematics has observed: the students always have to be taught what they should have learned in the preceding course. (We, the teachers, were of course exceptions; it is consequently hard for us to understand the deficiencies of our students.) The average student does not really learn to add fractions in arithmetic class; but by the time he has survived a course in algebra he can add numerical fractions. He does not learn algebra in the algebra course; he learns it in calculus, when he is forced to use it. He does not learn calculus in the calculus course, either; but if he goes on to differential equations he may have a pretty good grasp of elementary calculus when he gets through. And so on through the hierarchy of courses; the most advanced course, naturally, is learned only by teaching it.

This is not just because each previous teacher did such a rotten job. It is because there is not time for enough practice on each new topic; and even if there were, it would be insufferably dull. Anybody who has really learned to interpolate in trigonometric tables can also interpolate in air navigation tables, or in tables of Bessel functions. He should learn, because interpolation is useful. But one cannot drill students on mere interpolation; not enough, anyway. So the students solve oblique triangles in order (among other things) to practice interpolation. One must not admit this to the students, but one may as well realize the facts.

Consequently, I claim that there is a place, and a use, even for nonsense like the solution of quartics by radicals, or Horner's method, or involutes and evolutes, or whatever your particular candidates for oblivion may be. Here are problems that might conceivably have to be solved; perhaps the methods are not the most practical ones; but that is not the point. The point is that in solving the problems the student gets practice in using the necessary mathematical tools, and gets it by doing something that has more motivation than mere drill. This is not the way to train mathematicians, but it is an excellent way to train mathematical technicians. Now we can understand why calculus improves the line officer. He needs to practice very simple kinds of mathematics; he gets this practice in less distasteful form by studying more advanced mathematics.

It is the fashion to deprecate puzzle problems and artificial story problems. I think that there is a place for them too. Problems about mixing chemicals or sharing work, however unrealistic, give good practice and even have a good deal of popular appeal: witness the frequency with which puzzle problems appear in newspapers, magazines, and the flyers that come with the telephone bill. There was once a story in *The Saturday Evening Post* whose plot turned on the interest aroused by a perfectly preposterous diophantine problem about sailors, coconuts, and a monkey. It is absurd to claim that only "real" applications should be used to illustrate mathematical principles. Most of the real applications are too difficult and/or involve too many side issues. One begins the study of French with simple artificial sentences, not with the philosophical writings of M. Sartre. Similarly one has to begin the study of a branch of mathematics with simple artificial problems.

We may dislike this state of affairs, but as long as it exists we must face it. It would be pleasant to teach only the new and exciting kinds of mathematics; it would be comforting to teach only the really useful kinds. The traditional topics are some of the topics that once were either new and exciting, or useful. They have persisted partly by mere inertia—and that is bad—but partly because they still serve a real purpose, even if it is not their ostensible purpose. Let us keep this in mind when we are revising the curriculum.

CALCULUS AS AN EXPERIMENTAL SCIENCE*†

I hope that my title was not too misleading. I am not going to suggest that calculus should somehow be based on experiment, but rather that calculus should be presented to the student in the same spirit as the experimental sciences. The point that I hope to make is, briefly, that proofs are to mathematics what experiments are to physics (or chemistry, or biology), and that our teaching can profit by the analogy.

Let us first of all be clear about what calculus is. There are two big ideas, the derivative and the integral. Geometrically, these are the slope of a curve and the area under a curve. Of course they frequently, even usually, appear in nongeometrical forms: a derivative might represent a mass density and an integral might represent work, for instance. However, translating to and from geometry should not bother anyone who has ever done something like drawing a vector diagram of forces. The predecessors of Newton and Leibnitz knew perfectly well how to determine tangents and areas, but they had to approach each problem from first principles. The great contribution of Newton and Leibnitz was precisely to make the procedures for finding tangents, areas, etc. into a calculus, that is, a systematic way of calculating—a collection of algorithms, to use the currently fashionable word. Moreover, they didn't really understand—in the modern sense—why the algorithms worked. Perhaps it will make the point clearer if I use a very elementary example. If you write two numbers with Roman numerals, and want to multiply them, you can work out

---

*Vice-presidential address given at the meeting of the American Association for the Advancement of Science, December 28, 1970, as part of a Symposium on Mathematics in the Undergraduate Program in the Sciences, jointly sponsored by CUPM. The opinions expressed are those of the author, and are not to be taken as representing CUPM policy.

†*Amer. Math. Monthly* **78** (1971), 664–667; reprinted in *Two-Year College Math. J.* **2**, no. 1 (1971), 36–39.

the product if you understand the commutative, associative, and distributive laws of multiplication and addition for integers, but it takes time. Presumably the Romans used some other method, perhaps some kind of abacus (a simple digital computer). A more convenient way is to use Arabic numerals and the rules for manipulating them—a kind of calculus, in fact. In either case we have substituted rules or procedures for thinking about what is actually going on. Note incidentally that one can often use a calculus successfully without fully understanding why it works. (Does a digital computer understand arithmetic?)

Once invented, differential and integral calculus were very successful at solving certain kinds of geometrical problems, and hence physical problems that can be represented geometrically, and hence problems in physics, chemistry, economics, etc., even when they are not represented geometrically. Consequently calculus became the standard language for talking about the subjects in which it was most successful, it has remained so to this day, and seems likely to continue for some time into the future. This is why every scientist has to study calculus, although he often wonders at first why he should have to. Another way of saying much the same thing is that calculus is used because it facilitates the study of models of observed phenomena. If a biological process, for instance, can be modeled as a differential equation, calculus can take over and predict properties of the process without using any biological thought, and the biologist can then compare the prediction with experiment—in this way he may save considerable time and thought. It is hard to get this idea across to the beginning student, especially when he doesn't know any biology yet.

Everybody admits, I suppose, that the sciences other than mathematics are based on experiment. Things that can be checked by experiment are accepted: things that disagree with experiment are not. However, I am not aware of any physics or chemistry or biology course that repeats all the classical experiments, or even any of those that are particularly difficult or time-consuming. At least in the science courses I took (of course, this was a long time ago) we were told that certain things had been established experimentally, and maybe (not always) what the experiment was like. Inspection of some current textbooks suggests that things haven't changed much in 40 years.

Now I do not know any experimental scientists who seem to feel uncomfortable about this state of affairs, although for all I know they may worry about it in secret. Nor do they seem to worry about the necessity of sometimes giving oversimplified or even mildly fallacious reasons why the experiment comes out as it does. For example, why does an airplane stay up? Elementary texts give theoretical reasons that do not seem very convincing; the real theoretical reasons are clearly too sophisticated for elementary courses; it presumably would be possible also to rely on experimental measurements of the flow around a wing, but few, if any, physics courses bring wind-tunnels into the classroom. Similarly, first-year physics courses usually teach Newtonian mechanics, rather than relativistic mechanics or quantum mechanics. In calling attention to this, I do not intend to criticize the current teaching of the experimental sciences; in fact, I do not see what else could be done, and indeed I want to use the experimental scientists' approach as a model.

Mathematics is not at all an experimental science, but there is a rather exact parallel between mathematics and the experimental sciences. In mathematics we believe things, not because we did an experiment, but because we proved them. At least—and this continues the analogy—we believe things because *somebody* proved them; we have not necessarily studied the proof ourselves. Thus the proof is to mathematics as the experiment is to physics, chemistry, biology, and so on. Perhaps we are better off than the experimental scientist in one respect—we know that we *could* read the proof if we tried, whereas the experiment may be too difficult or too expensive for the experimental scientist to hope to repeat for himself or his class. Of course we want our students to believe what we say because we have done something that really carries conviction. What shall we do? The answer that most mathematicians believe in, or profess to believe in, or act as if they believed in, is that they ought to present their students with formal proofs of everything that they tell them. The effect of this is to make calculus a chapter in the theory of functions of a real variable. There are several reasons for this attitude. There are mathematicians who didn't understand calculus themselves to begin with, but now do; and, filled with missionary zeal, wish to spread the light. There are those who feel that it is intellectually dishonest not to tell everything that they know. There are those who feel that anything less than full explanations cheats the student. And there are those who understand the proofs but can't solve the problems: theory is always easier than technique. In any case, only just so much time is available. In order to make the best use of it, I claim that the teacher of calculus would do well to follow the lead of the experimental scientist: let him give proofs when they are easy and justify unexpected things; let him omit tedious or difficult proofs, especially those of plausible things. Let him give easy proofs under simplified assumptions rather than complicated proofs under general hypotheses. Let him by all means always give correct statements, but not necessarily the most general ones that he knows.

Let us see how these principles apply to some topics in calculus. (1) One of the unexpected results is the formula for the derivative of a product. Most beginners will guess it wrong. The proof is easy and completely convincing. One should by all means give it. (2) A function with a positive derivative is increasing. This looks, but isn't, tautological; the point at issue, generating a global property from a local one, is rather subtle. The proof is not illuminating, and might well be skipped. (3) It is certainly necessary to define the definite integral, but to prove that the integral of a continuous function exists is both technically demanding and time-consuming. This seems to be a clear case for "it can be proved." (4) The uniqueness theorem for solutions of a second-order linear differential equation is only too plausible—don't the initial position and velocity determine the motion? The proof is time-consuming, but the facts are easy to state precisely and meaningfully. (5) Assuming that Fourier series get into the calculus (they usually don't), it would be difficult, time-consuming, and unconvincing to *prove* any really useful convergence theorem. On the other hand, should Fourier series be left out just because we cannot prove a satisfying theorem about them? It is easy enough to state one, and there is no excuse for stating an incorrect one.

I am going to be accused by my colleagues of advocating a cookbook approach to calculus. This I deny. There was once a really cookbook approach to calculus, in which the student had to listen to incomprehensible nonsense until he developed a sound intuition (if he ever did). The approach that some of my colleagues favor makes the student listen to incomprehensible sense instead. I think the experimental scientists do better. Let me illustrate the difference with an example. A cookbook approach to maximum and minimum problems leads the student to approach all problems by setting a derivative equal to zero and testing by the sign of the second derivative. This traditional procedure can, in fact, lead to mistakes. A rigorous approach demands a long series of preliminary theorems about maxima, mean values and derivatives. What I prefer is something like this: observe that if there is a maximum where there is a derivative, the derivative must be zero; then the maximum must be at one of the (usually small number of) points where the derivative is zero or doesn't exist, or else at an endpoint. A small amount of computation will usually decide; and we avoid the second-derivative test, which in spite of its theoretical elegance is usually quite impractical. It seems to me that an approach of this kind is very much in the spirit in which experimental sciences are usually presented; and in practice it seems to give the students more capability with calculus, and sooner, than the theorem-proving approach that has been so popular.

CAN WE MAKE MATHEMATICS INTELLIGIBLE?*

Why is it that we mathematicians have such a hard time making ourselves understood? Many people have negative feelings about mathematics, which they blame, rightly or wrongly, on their teachers [1]. Students complain that they cannot understand their textbooks; they have been doing this ever since I was a student, and presumably for much longer than that. Professionals in other disciplines feel compelled to write their own accounts of the mathematics they had trouble with. However, it was not until after I became editor of this Monthly that I quite realized how hard it is for mathematicians to write so as to be understood even by other mathematicians (outside of fellow specialists). The number of manuscripts rejected, not for mathematical deficiencies but for general lack of intelligibility, has been shocking. One of my predecessors had much the same experience 35 years earlier [2].

To put it another way, why do we speak and write about mathematics in ways that interfere so dramatically with what we ostensibly want to accomplish? I wish I knew. However, I can at least point out some principles that are frequently violated by teachers and authors. Perhaps they are violated because they contradict what many of my contemporaries seem to consider to be self-evident truths. (They also have little in common with the MAA report on how to teach mathematics [3].)

ABSTRACT DEFINITIONS

Suppose you want to teach the "cat" concept to a very young child. Do you explain that a cat is a relatively small, primarily carnivorous mammal with retractile claws,

---

*Amer. Math. Monthly **88** (1981), 727–731.

a distinctive sonic output, etc.? I'll bet not. You probably show the kid a lot of different cats, saying "kitty" each time, until it gets the idea. To put it more generally, generalizations are best made by abstraction from experience. They should come one at a time; too many at once overload the circuits.

There is a test for identifying some of the future professional mathematicians at an early age. These are students who instantly comprehend a sentence beginning "Let X be an ordered quintuple $(a, T, \pi, \sigma, \mathcal{B})$, where…". They are even more promising if they add, "I never really understood it before." Not all professional mathematicians are like this, of course; but you can hardly succeed in becoming a professional unless you can at least understand this kind of writing.

However, unless you are extraordinarily lucky, most of your audience will not be professional mathematicians, will have no intention of becoming professional mathematicians, and will never become professional mathematicians. To begin with, they won't understand anything that starts off with an abstract definition (let alone with a dozen at once), because they don't yet have anything to generalize from. Please don't immediately write me angry letters explaining how important abstraction and generalization are for the development of mathematics: I *know* that. I also am sure that when Banach wrote down the axioms for a Banach space he had a lot of specific spaces in mind as models. Besides, I am discussing only the communication of mathematics, not its creation.

For example, if you are going to explain to an average class how to find the distance from a point to a plane, you should first find the distance from $(2, -3, 1)$ to $x - 2y - 4z + 7 = 0$. After that, the general procedure will be almost obvious. Textbooks used to be written that way. It is a good general principle that, if you have made your presentation twice as concrete as you think you should, you have made it at most half as concrete as you ought to.

Remember that *you* have been associating with mathematicians for years and years. By this time you probably not only think like a mathematician but imagine that everybody thinks like a mathematician. Any nonmathematician can tell you differently.

ANALOGY

Sometimes your audience will understand a new concept better if you explain that it is similar to a more familiar concept. Sometimes this device is a flop. It depends on how well the audience understands the analogous thing. An integral is a limit of a sum; therefore, since sums are simpler (no limiting processes!), students will understand how integrals behave by analogy with how sums behave. Won't they? In practice, they don't seem to. Integrals are simpler than sums for many people, and there may be some deep reason for this [4].

VOCABULARY

Never introduce terminology unnecessarily [5]. If you are going to have to mention a countable intersection of open sets—just once!—there is no justification for defining G_δ's and F_σ's.

I have been assured that nobody can really understand systems of linear equations without all the special terminology of modern linear algebra. If you believe this you must have forgotten that people understood systems of linear equations quite well for many years before the modern terminology had been invented. The terminology allows concise statements; but concision is not the alpha and omega of clear exposition. Modern terminology also lets one say more than could be said in old-fashioned presentations. Nevertheless, at the beginning of the subject a lot of the students' effort has to go into memorizing *words* when it could more advantageously go into learning mathematics. Paying more attention to vocabulary than to content obscures the content. This is what leads some students to think that the real difference between Riemann and Lebesgue integration is that in one case you divide up the x-axis and in the other you divide up the y-axis.

If you think you can invent better words than those currently in use, you are undoubtedly right. However, you are rather unlikely to get many people except your own students to accept your terminology; and it is unkind to make it hard for your students to understand anyone else's writing. One Bourbaki per century produces about all the neologisms that the mathematical community can absorb.

In any case, if you *must* create new words, you can at least take the trouble to verify that they are not already in use with different meanings. It has not helped communication that "distribution" now means different things in probability and in functional analysis. On the other hand, if you need to use old but unfashionable words it is a good idea to explain what they mean. A friend of mine was rebuked by a naïve referee for "inventing" bizarre words that had actually been invented by Kepler.

It is especially dangerous to assume either that the audience understands your vocabulary already or that the words mean the same to everybody else that they do to you. I know someone who thinks that everybody from high school on up knows all about Fourier transforms, in spite of considerable evidence to the contrary. Other people think that everybody knows what they mean by Abel's theorem, and therefore never say which of Abel's many theorems they are appealing to.

An even more serious problem comes from what (if it didn't violate my principles) I would call geratologisms: that is, words and phrases that, if not actually obsolete in ordinary discourse, are becoming so. Contemporary prose style is simpler and more direct than the style of the nineteenth century—except in textbooks of mathematics. While I was writing this article I was teaching from a calculus book that begins a problem with, "The strength of a beam varies directly as...". I do not know whether the jargon of variation is still used in high schools, but in any case it isn't learned: only one student in a class of 45 had any idea what the book meant (and he was a foreigner). Blame the students if you will, blame the high schools; for my own part I blame the authors of the textbook for not realizing that contemporary students speak a different language. Another current calculus book says, "Particulate matter concentrations in parts per million theoretically decrease by an inverse square law." You couldn't get away with that in *Newsweek* or even in *The New Yorker*, but in a textbook. . . .

Authors of textbooks (lecturers, too) need to remember that they are supposed to be addressing the students, not the teachers. What is a function? The textbook wants you to say something like, "a rule which associates to each real number a uniquely specified real number," which certainly defines a function—but hardly in a way that students will comprehend. The point that "a definition is satisfactory only if the students understand it" was already made by Poincaré [6] in 1909, but teachers of mathematics seem not to have paid much attention to it.

The difficulties of a vocabulary are not peculiar to mathematics; similar difficulties are what makes it so frustrating to try to talk to physicians or lawyers. They too insist on a rich technical language because "it is so much more precise that way." So it is, but the refined terminology is clearer only when rigorous distinctions are absolutely necessary. There is no use in emphasizing refined distinctions until the audience knows enough to see that they are needed.

SYMBOLISM

Symbolism is a special kind of terminology. Mathematics can't get along without it. A good deal of progress has depended on the invention of appropriate symbolism. But let's not become so fascinated by the symbols that we forget what they stand for. Our audience (whether it is listening or reading) is going to be less familiar with the symbolism than we are. Hence it is not a good idea (to take a simple example) to say "Let f belong to L^2" instead of "Let f be a measurable function whose square is integrable," unless you are sure that the audience already understands the symbolism. Moreover, if you are not actually going to use L^2 as a Hilbert space, but want only the properties of its elements as functions, the structure of the space is irrelevant and calling attention to it is a form of showing off—mild, but it *is* showing off. If the audience doesn't know the symbolism, it is mystified; if it does know, it will be wondering when you are going to get to the point.

My advice about new terminology applies with even greater force to new symbolism. Do not create new symbolism, or change the old, unnecessarily; and admit (if necessary) that usage varies and explain the existing equivalences. If your $\Phi(x)$ also appears in the literature as $P(x)$ or $P(x) + \frac{1}{2}$ or $F(x)$, *say so*. Irresponsible improvements in notation have already caused enough trouble. I don't know who first thought of using θ in spherical coordinates to mean azimuth instead of colatitude, as it almost universally did and still does in physics and in advanced mathematics. It's superficially a reasonable convention because it makes θ the same as in plane polar coordinates; however, since r is different anyway, that isn't much help. The result is that students who go beyond calculus have to learn all the formulas over again. Such complications don't bother the true-blue pure mathematicians, those who would just as soon see Newton's second law of motion stated as $\mathbf{v} = (d/d\sigma)(\mathcal{R}\mathbf{q})$, but they do bother many students, besides irritating physical scientists.

PROOFS

Only professional mathematicians learn anything from proofs. Other people learn from explanations. I'm not sure that even mathematicians learn much from proofs in fields with which they are not familiar. A great deal can be accomplished with arguments that fall short of being formal proofs. I have known a professor (I hesitate to say "teacher") to spend an entire semester on a proof of Cauchy's integral theorem under very general hypotheses. A collection of special cases and examples would have carried more conviction and left time for more varied and interesting material, besides leaving the audience better equipped to understand, apply, generalize, and teach Cauchy's theorem.

I cannot remember who first remarked that a sweater is what a child puts on when its parent feels cold; but a proof is what students have to listen to when the teacher feels shaky about a theorem. It has been claimed [7] that "some of the most important results... are so surprising at first sight that nothing short of a proof can make them credible." There are fewer of these than you think.

Experienced parents realize that when a child says "Why?" it doesn't necessarily want to hear a reason; it just wants more conversation. The same principle applies when a class asks for a proof.

RIGOR

This is often confused with generality or completeness. In spite of what reviewers are likely to say, there is nothing unrigorous in stating a special case of a theorem instead of the most general case you know, or a simple sufficient condition rather than a complicated one. For example, I prefer to give beginners Dirichlet's test for the convergence of a Fourier series: "piecewise monotonic and bounded" is more comprehensible than "bounded variation"; and, in fact, equally useful after one more theorem (learned later).

The compulsion to tell everything you know is one of the worst enemies of effective communication. We mathematicians would get along better with the Physics Department if, for example, we could bring ourselves to admit that, although their students need some Fourier analysis for quantum mechanics, they don't need a whole semester's worth—two weeks is nearer the mark.

Being more thorough than necessary is closely allied to **pedantry**, which (my dictionary says) is "excessive emphasis of trivial details."

Here's an example. Suppose students are looking for a local minimum of a differentiable function f, and they find critical points at $x = 2$, $x = 5$, and nowhere else. Suppose also that they do not want to use (or are told not to use) the second derivative. Some textbooks will tell them to check $f(2 + h)$ and $f(2 - h)$ for all small h. Students naturally prefer to check $f(3)$ and $f(1)$. The pedantic teacher says, "No"; the honest teacher admits that any point up to the next critical point will do.

ENTHUSIASM

Teachers are often urged to show enthusiasm for their subjects. Did you ever have to listen to a really enthusiastic specialist holding forth on something that you did not know and did not want to know anything about, say the bronze coinage of Poldavia in the twelfth century or "the doctrine of the enclitic *De*" [8]? Well, then.

SKILLS

A great deal of the mathematics that many mathematicians support themselves by teaching consists of subjects like elementary algebra or calculus or numerical analysis—skills, in short. It is not always easy to tell whether a student has acquired a skill or, as we like to put it, "really" learned a subject. The difficulty is much like that of deciding whether apes can use language in a linguistically interesting way or whether they have just become very clever at pushing buttons and waving their hands [9]. Mathematical skills are like any other kind. If you are learning to play the piano, you usually start by practicing under supervision; you don't begin with theoretical lectures on acoustical vibrations and the internal structure of the instrument. Similarly for mathematical skills. We often read or hear arguments about the relative merits of lectures and discussions, as if these were the only two ways to conduct a class. Having students practice under supervision is another and very effective way. Unfortunately it is both untraditional and expensive.

Even research in mathematics is, to a considerable extent, a teachable skill. A student of G. H. Hardy's once described to me how it was done. If you were a student of Hardy's, he gave you a problem that he was sure you could solve. You solved it. Then he asked you to generalize it in a specific way. You did that. Then he suggested another generalization, and so on. After a certain number of iterations, you were finding (and solving) your own problems. You didn't necessarily learn to be a second Gauss that way, but you could learn to do useful work.

LECTURES

These are great for arousing the emotions. As a means of instruction, they ought to have become obsolete when the printing press was invented. We had a second chance when the Xerox machine was invented, but we seem to have muffed it. If you *have* to lecture, you can at least hand out copies of what you said (or wish you had said). I know mathematicians who contend that only through their lectures can they communicate their personal attitudes toward their subjects. This may be true at an advanced level, for pre-professional students. Otherwise I wonder whether these mathematicians' personalities are really worth learning about, and (if so) whether the students couldn't learn them better some other way (over coffee in the cafeteria, for example.)

One of the great mysteries is: How can people manage to extract useful information from incomprehensible nonsense? In fact, we can and do. Read, for example, in Morris Kline's book [10] about the history of the teaching of calculus. Perhaps this talent that we have can explain the popularity of lectures. One incomprehensible lecture is not enough, but a whole course may be effective in a way that one incomprehensible book never can. I still contend that a comprehensible book is even better.

CONCLUSION

I used to advise neophyte teachers: "Think of what your teachers did that you particularly disliked—and don't do it." This was good advice as far as it went, but it didn't go far enough. My tentative answer to the question in my title is, "Yes; but don't be guided by introspection." You cannot expect to communicate effectively (whether in the classroom or in writing) unless and until you understand your audience. This is not an easy lesson to learn.

References

[1] See, for example, Sydney J. Harris, column for February 9, 1980, Chicago Sun-Times and elsewhere.

[2] L. R. Ford, Retrospect, this *Monthly*, 53 (1946) 582–585.

[3] College Mathematics: Suggestions on How to Teach It, Mathematical Association of America, 1972.

[4] D. R. Stoutemyer, Symbolic computation comes of age, SIAM News, 12, no. 6 (December 1979) 1, 2, 9.

[5] The same point has been made by P. R. Halmos in How to Write Mathematics, L'Enseignement mathématique, (2) 16 (1970) 123–152.

[6] H. Poincaré, Science et méthode, 1909, Book II, Chapter 2.

[7] H. and B. S. Jeffreys, Methods of Mathematical Physics, 2nd ed., Cambridge University Press, 1950, p. v.

[8] Robert Browning, "A Grammarian's Funeral," in The Complete Poetic and Dramatic Works of Robert Browning, Houghton Mifflin, Boston and New York, 1895, pp. 279–280.

[9] For example, E. S. Savage-Rumbaugh, D. M. Rumbaugh, and S. Boysen, Do apes use language? Amer. Scientist, 68 (1980) 49–61.

[10] Morris Kline, Mathematics: The Loss of Certainty, Oxford University Press, New York, 1980.

BOXING THE CHAIN RULE*

If you teach calculus, you know how much trouble students have with the chain rule for differentiation. I should like to share with you a teaching technique for the chain rule that seems not to be well known—at least, I haven't seen it in any textbook (but of course I haven't by any means seen all the textbooks!) My wife and I have used this technique successfully for many years; I think she invented it, but after forty years it is hard to be sure.

To begin with, I tell the students that the derivative of $\square^n$ is $n\square^{n-1}$ times the derivative of $\square$ (using whatever symbolism for "the derivative of" that the current textbook prefers.) Orally, "The derivative of box to the nth power is n times box to the $n - 1$ times the derivative of box." I tell them that they can put whatever they like in the boxes provided that they put the same thing in all three of them. So, for instance,

$$\frac{d}{dx}(2x^2 - x)^n = n(2x^2 - x)^{n-1}\frac{d}{dx}(2x^2 - x).$$

I usually emphasize the arbitrariness of what goes into the box (and get a cheap laugh) by putting in "Reagan" or a sketch of a bird or whatever occurs to me.

Then of course, more generally,

$$\frac{d}{dx}f(\square) = f'(\square)\frac{d}{dx}(\square)$$

**California MathematiCs* 7, no. 1 (1982), 36. Reprinted by permission.

and there's the chain rule. ("The derivative of f of box is f' of box times the derivative of box.") It's amazing how much more easily this is apprehended, than even

$$\frac{d}{dx}f(\ ) = f'(\ )\frac{d}{dx}(\ ).$$

Students who learn the rule this way just don't forget the extra derivative. I usually claim that the box version was the original one and the form with parentheses was introduced by a lazy printer, which sounds plausible but is probably historically incorrect.

NAMES OF FUNCTIONS: THE PROBLEMS OF TRYING FOR PRECISION*

TO STUDENTS

Are you being confused by a textbook or a teacher who insists that $f(x)$ is the value of a function at the point x, whereas f is the name of the function? I wouldn't be surprised if you were, especially if your text goes on (most of them do) to present formulas like $\frac{d}{dx}\sin x = \cos x$ or $\frac{d}{dx}x^2 = 2x$, apparently without reflecting that, if $\sin x$ is the value of the sine function at x, then $\frac{d}{dx}\sin x$ is, strictly speaking, a meaningless formula. (We differentiate functions, not numbers.) You could get around this particular problem by writing $\frac{d}{dx}\sin = \cos$; but what *are* you to do about the function that has the value x^2 at x? You can't very well write ^{2} as the name of the function. What most people—you (probably), your teacher (quite likely), and your pocket calculator (almost certainly)—want to call the function is x^2, but this would violate the teacher's principles. I want to propose a simple way out of this bind.

First let's recall that there is one kind of function that has a well-established and unambiguous notation already: a sequence. The sequence $\{n^2\}_1^\infty$ is the function whose value at n is n^2, the domain being understood to be the set of positive integers. If you need a different domain you can write symbols like $\{n^2\}_3^\infty$ or $\{n^2\}_{10}^{100}$ or $\{n^2\}_{\text{odd } n}$. The identity sequence is $\{n\}_1^\infty$. In other words, we know a sequence

Math. Mag.* **56 (1983), 175–176.

when we meet one because it comes with its domain and its value at each domain point—which is just what every definition of a function is supposed to provide.

Why then don't we denote the function that has the value x^2 at the real number x by $\{x^2\}_R$ or $\{x^2\}_1^\infty$, and so on? The identity is $\{x\}$, and $\frac{d}{dx}\{x\} = \{1\}$, as it should be. This would be unambiguous, compact, and would require no special symbols to be learned.

If you use any notation a great deal you tend to abbreviate it, just as the word "radix" (meaning "root") got cut down to the modern square root sign. If your usual domain for functions is all real numbers, you will probably just write $\{\sin x\}$ instead of $\{\sin x\}_{-\infty<x<\infty}$. After doing that for a while, you will find yourself dropping the braces too—and there you will be with functions named x^2, $\sin x$, and so on, just as the keys on the calculator, or the tables in the textbook say. The difference is that you now ought to be able, on demand, to explain the difference between $\sin x$, meaning a number, and $\sin x$, meaning a function—which is probably what your teacher was hoping for all along.

TO TEACHERS

Are you quite, quite sure that when you make students learn that f is a function and $f(x)$ is a value of a function, they are really learning what functions and values are? Or are they just parroting words? I've seen plenty of students who could give you a letter-perfect definition of a derivative but were helpless if you asked them what $\lim_{x\to\pi} \frac{\sin x}{x-\pi}$ is. How many of them can tell you what is the sine of the angle whose sine is x?

Maybe you deplore the calculator people's putting "x^2" on the squaring key. Your most impassioned arguments aren't going to stop them. "If you can't lick 'em, join 'em."

Abbreviations are an important mathematical tool. If we weren't allowed to use them, we'd still be writing $\lim\limits_{y\to x} \frac{f(y)-f(x)}{y-x}$ instead of $f'(x)$. Bourbaki calls this sort of thing "abus de langage;" ordinary people call it shorthand. Admittedly it has disadvantages. The notation $\sin^2 x$ (Gauss is on record as detesting it) is shorthand for $(\sin x)^2$. The notation $\log^2 x$ can mean either $(\log x)^2$ or $\log(\log(x))$. However, people seem to prefer ambiguous notations to cumbersome ones.

In integration theory, a Lebesgue integrable function is not a function at all; it is an equivalence class of functions. Do we indicate this by our notation? Not that I've ever noticed.

I have heard a linguist claim that the eccentric orthography of English serves a useful purpose besides making it possible to have spelling bees: it helps us pick out the correct meanings of words that we see. Perhaps it's just as well that we *don't* use strictly consistent notations.

DOES "HOLDS WATER" HOLD WATER?*

A number of calculus books give the mnemonic that a curve which is concave upward "holds water," whereas one which is concave downward "spills water." Recently, a student asked one of my colleagues why the graph of $|x|^{1/2}$ is not concave upward in an interval containing 0, because it would evidently hold water. Indeed, there are actual glasses that have a similar shape (with the addition of a stem) and do hold wine (if not water). The problem arises because the mnemonic, like most mnemonics, is flawed: "holding water" is neither necessary nor sufficient for a given curve to be concave upward. The student's example can be augmented by $y = -|x|^{1/2}$, which clearly "spills water."

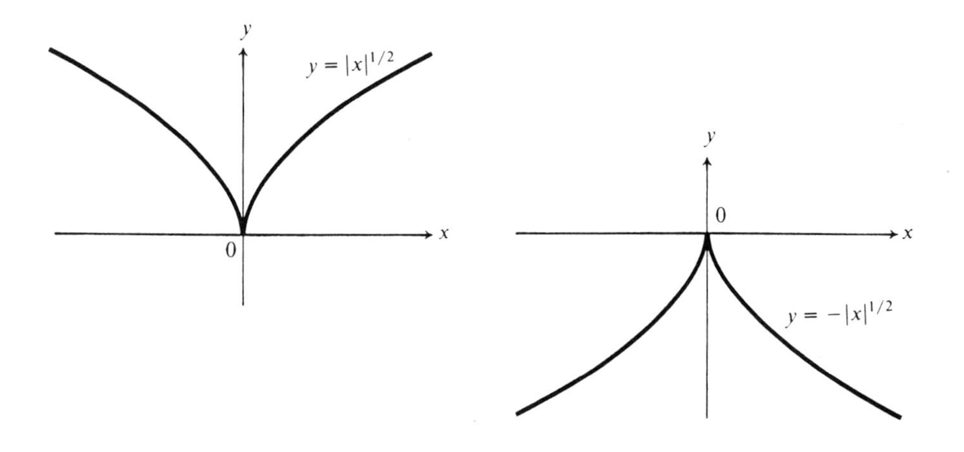

College Math. J. **17** (1986), 341.

The late Professor W. R. Ransom used to use "bowl shaped" and "dome shaped." I have often followed his lead without realizing that this mnemonic is equally flawed. Wide bowls with sides that curve downward are not uncommon, and the domes on Eastern Orthodox churches are not concave downward at the extreme top. I have also seen the mnemonic that a smile is concave up and a frown (or scowl) is concave down; this connects nicely with the colloquial use of "up" and "down." Here smiles are to be thought of as in cartoons and graffiti; actual smiles can be misleading, especially when they are crooked.

I am indebted to A. M. Trembinska for telling me about the puzzled student.

SECTION 14

POLYNOMIALS

EXTREMAL PROBLEMS FOR POLYNOMIALS*

One of the most romantic of all story plots is the one about the inventor who patiently works for years in obscurity and finally achieves resounding success. Such stories are rare in mathematics; this is a report on one of them.

So—once upon a time (just about a century ago) the chemist Mendeleev (the one who invented the periodic table of the elements) was interested in the relationship between the specific gravity of a solution and the concentration of the dissolved substance. This question seems no longer to be of much interest to physical chemists, but a service station still uses the relationship to check the concentration of antifreeze in the cooling system of your automobile, and a brewer or vintner uses it to test the concentration of alcohol in beer or wine. Mendeleev made accurate graphs (they agree with modern tables to three significant figures) for many substances, and found that the curves exhibited small kinks, so that no very simple formulas would represent them. He proceeded to fit the curves by a succession of parabolic arcs, and found that the arcs did not join smoothly. He then wondered whether the corners corresponded to genuine phenomena or were just the result of errors of measurement. He seems to have reasoned somewhat as follows. Suppose I have two quadratic polynomials $P_2(x)$ and $Q_2(x)$ on adjacent intervals, with graphs that join continuously but at an angle. Suppose I know that the slope of P_2 on its interval cannot exceed a number S, whereas the slope of Q_2 is distinctly larger than S; then I cannot expect to replace P_2 and Q_2 by a single quadratic on the union of the intervals. Hence Mendeleev was led to ask, if we know how large a quadratic polynomial P_2 is on a given interval $[a, b]$, how large can the derivative P_2' be on the same interval?

*Amer. Math. Monthly **85** (1978), 473–475.

Mendeleev was enough of a mathematician to answer the question for himself, the answer being that if $|P_2(x)| \le M$ on an interval of length $2L$, then $|P_2'(x)| \le 4ML$ on the same interval, 4 being the best (that is, smallest) number that will always work. (Incidentally this led Mendeleev to conclude that the corners were genuine.) If you were a chemist and found a pretty theorem like this, you would want to tell it to a mathematician, and that is just what Mendeleev did—he told it to A. A. Markov. If you are a mathematician and a chemist tells you a nice result about quadratic polynomials, you naturally want to generalize it to polynomials of degree n, and that is just what Markov did: he proved that if $P_n(x)$ is of degree n and $|P_n(x)| \le M$ on an interval of length $2L$, then $|P_n'(x)| \le n^2ML$ on the same interval, and that n^2 is the best constant. This has become famous as Markov's theorem; it is not only elegant but has many applications very different from the one that interested Mendeleev.

As it happens, equality can be attained in Markov's theorem only at the end-points of the interval in question. This suggests the problem of finding the point-by-point maximum for $|P_n'(x)|$: to find $M_n(x) = \max |P_n'(x)|$, where the maximum is taken over all polynomials of degree n such that $|P_n(x)| \le M$ on $[a, b]$. Until recently not much was known about $M_n(x)$ except for what S. Bernstein proved in 1912. This is most easily stated for the interval $[-1, 1]$: if $|P_n(x)| \le M$ on $[-1, 1]$ then $|P_n'(x)| \le n(1 - x^2)^{-1/2}$ on the same interval. This is useful in many applications, but is of course much weaker than Markov's theorem for x near ± 1; on the other hand it is much better for x near 0. Very many papers have been written about Bernstein's and Markov's theorems and similar extremal problems, but the problem of finding $M_n(x)$ remained untouched for many years.

The last statement is not quite true. Markov's paper (written in Russian in a not very accessible journal) must be one of the most often cited papers—everybody who writes about this kind of problem feels compelled to cite it—and one of the least read. If you go to the trouble to read it you will find that Markov not only raised the problem that I just mentioned, but took a considerable step toward solving it, and did solve it for $n = 2$.

With this much background we can proceed to the story. Around 1930 E. V. Voronovskaya began to publish a series of short articles dealing with the Hausdorff moment problem, which is the study of sequences $\mu_n = \int_0^1 t^n \, dH(t)$, where H is a function of bounded variation. These articles are quite technical, use a rather forbidding special terminology, give few proofs, and attracted almost no attention. Very likely nobody who read them realized what they were leading up to. But—a quarter-century later Voronovskaya's work really paid off. She could solve not only the point-by-point Markov problem, but also a great variety of other extremal problems that had previously seemed too difficult for anyone to do anything with. Her work showed in particular that the kind of solution that the world was used to seeing was too simple to be possible for anything but the simplest problems. The only full account of her work appeared in 1963 in a book published by the Leningrad Institute for Electrotechnical Communication (and not easy to come by), although some of the details, including the solution of Markov's problem, had come out in journal articles in the late 1950's; an English translation [1] of the book was published in 1970.

How does the Hausdorff moment problem connect with extremal problems for polynomials? Answer: via the Riesz representation theorem for continuous linear functions on the space C of continuous functions $X(t)$ on $[0,1]$. This theorem says that these functionals are of the form $F(X) = \int_0^1 X(t)\,dH(t)$ with H of bounded variation, and moreover the norm of the functional F is the total variation of H. Now since polynomials form a dense set in C, a linear functional F is completely determined by its values on the polynomials or, since F is linear, by its values on $X_n(t) = t^n$, $n = 0, 1, 2, \ldots$; that is, by the moments $\int_0^1 t^n dH(t)$. Furthermore, $|F(P_n(t))| \le \|F\| \max_t |P_n(t)|$, with $\|F\| = \operatorname{var} H = \sup |F(P_n(t))|$ for $\|P_n(t)\| \le 1$. Now let $F(P_n) = P_n'(x)$ for a particular value of x. This functional corresponds to the moment sequence $0, 1, 2x, 3x^2, \ldots$; the norm of F is $\max |P_n'(x)|$ over all polynomials of degree n for which $\max_t |P_n(t)| \le 1$, and Markov's general problem is seen to be the problem of determining the norm of this particular functional. This reinterpretation does not, of course, solve the problem, but it does suggest a systematic attack on a wide variety of such problems in place of the many ad hoc methods that have been devised for special classes of extremal problems.

The actual solution of Markov's problem is not easy to describe. The extremals of the original Markov problem—to find $\max |P_n'(x)|$ given $\max |P_n(x)|$, on (say) the interval $[0,1]$—are the Čebyšev polynomials $T_n(x) = \cos n \cos^{-1}(2x-1)$. It turns out that the interval $[0,1]$ breaks up into a succession of three kinds of intervals. In intervals of the first kind, $|P_n'(x)|$ is maximized by $|T_n'(x)|$. In another set of intervals the extremal polynomials have the form $\pm T_n(\lambda x)$ or $\pm T_n(\lambda(1-x))$. But in the remaining intervals the extremal polynomials are the much less well known Zolotarev polynomials, indeed different ones at each point, and the maximal function $M_n(x)$ has only been exhibited parametrically, although it can be evaluated numerically by computer. Reference [1] shows graphs for $n = 2$ (where the Zolotarev polynomials do not occur), 3, 4, and 5. For $n > 2$ the function $M_n(x)$ is, perhaps surprisingly, neither monotonic nor convex; it is, however, continuously differentiable (which shows that Markov's $M_3(x)$ cannot be correct for all values of x). Perhaps one should not be too surprised at finding so much complexity in an apparently simple problem, especially since one of Voronovskaya's theorems (discovered independently by Rogosinski) is that *every* polynomial is extremal for *some* extremal problem.

Reference

[1] E. V. Voronovskaya, The Functional method and its applications, American Mathematical Society, 1970.

PERIODIC ENTIRE FUNCTIONS*

An entire function $f(z)$ is said to be of exponential type if $|f(z)| \leq Ae^{B|z|}$ for some numbers A and B. Evidently e^{ikz} is of exponential type and so is any trigonometric polynomial $\sum_{-n}^{n} a_k e^{ikz}$. A trigonometric polynomial is, in addition, periodic with period 2π. It is natural to expect that the converse holds: an entire function of exponential type which has period 2π is necessarily a trigonometric polynomial. This theorem has been discovered and proved a number of times (cf. [1] p. 109). The following very short proof reduces the theorem to the level of a classroom exercise.

Let $|f(z)| \leq Ae^{B|z|}$ and let $f(z)$ have period 2π. Then $g(w) = f(-i \log w)$ is uniform, and if n is an integer greater than B we have

$$w^n g(w) = \begin{cases} O(|w|^{2n}), & |w| \to \infty, \\ O(1), & |w| \to 0. \end{cases}$$

The second estimate shows that $w^n g(w)$ has a removable singularity at 0. Remove it. Then the first estimate shows that $w^n g(w)$ is a polynomial, so that $f(z) = g(e^{iz})$ is a trigonometric polynomial.

Reference

[1] R. P. Boas, Jr., Entire functions, Academic Press, New York, 1954.

*Amer. Math. Monthly **71** (1964), 782.

SECTION 15

LITERATURE AND MATHEMATICS

1724 LILLIPUTIANS*

In the first edition (1726) of *Gulliver's Travels*, at the end of Chapter 3 of the Voyage to Lilliput, Gulliver says that he was provided with food sufficient for 1724 Lilliputians. The second edition (1727) changes 1724 to 1728, which of course is correct on the scale of an inch to a foot. Some later editions have 1724 and some have 1728. The number 1724 is usually interpreted as a printer's error, but since it is mentioned several times it seems unlikely to have been merely a mistake in setting type. No correction is indicated in the large paper copy (now in the Victoria and Albert Museum) in which Charles Ford recorded Swift's corrections to the first edition. I suggest that 1724 is what Swift intended. The Lilliputians are petty in mind as well as in body, as Swift constantly emphasizes. It is quite in keeping with the satire for Gulliver to describe them as "most excellent mathematicians" and then show them unable to cube 12 correctly. It seems likely that Swift was the victim of a well-intentioned copy editor in 1727.

Amer. Notes and Queries* **6 (1968), 115–116. Reprinted by permission.

The following note on Shakespeare's Twelfth Night was coauthored with his father.

SHAKESPEARE'S TWELFTH NIGHT, II, iii, 25–27*

Sir Andrew. . . . I sent thee sixpence for thy leman. Had'st it?

Clown. I did impeticos thy gratillity, for Malvolio's nose is no whipstock. My lady has a white hand, and the Mermidons are no bottle-ale houses.

The editors of the New Cambridge Edition (p. 126) explain "impeticos" as "to pocket, in reference. . . to the professional long coat of the fool"; "gratillity" as "wretched little gratuity"; "whipstock" as "the type of stupidity and insentience" and "my lady" as "the girl who honors him with 'her white' hand."

These explanations are unsatisfactory and misleading for a number of reasons. If "impeticos" means "to pocket," then there is no point in the *for* since Malvolio's "stupidity and insentience" have nothing to do with the Clown's pocketing his tip. "Gratillity" cannot be "a wretched little gratuity" because sixpence in Shakespeare's time was the equivalent of half-a-crown to-day, a good tip even to the doorman at the Ritz. Moreover, if it was a "wretched little gratuity" it would not have paid for ale at a real inn with a sign. Besides, why did the clown have to take his girl to a really fine place?

If, however, one realizes that the Clown is punning on "petticoat" (cf. modern "skirt" as applied to a girl); that in his usual euphuistic manner he is calling his sixpence an "elegant little gratuity"; that he is conveying to his audience that he spent his tip all at one party because his girl (not being a red-handed servant girl)

The Explicator, **3**(4) (1945), no. 29. Reprinted with permission of Helen Dwight Reid Educational Foundation. Published by Heldref Publications, 1319 18th Street N.W., Washington D.C. 20036–1802.

demanded a high class treat, not a mere drink at a hedge ale house; and that he is hinting that Malvolio, who, as we find out later, hates "cakes and ale," would have had him whipped for his little affair if he could have; then the passage makes complete sense.

I suggest, therefore, the following paraphrase:

Sir Andrew. I sent you half a crown for your girl. Did you get it?

Clown. I spent your elegant little gratuity on my girl. I could get away with my little party because, though Malvolio sticks his long nose into everybody's business, he could not have me whipped for this affair. I spent it all because my girl costs money since she is no servant girl. Moreover I took her to a real inn with a sign, and not to one of those cheap places where they sell only bottled ale.

Ralph P. Boas, *Wheaton College* and
R. P. Boas, Jr., *Harvard University*

WHAT ST. AUGUSTINE DIDN'T SAY ABOUT MATHEMATICIANS*

At about the time when I was becoming seriously interested in mathematics, I had a friend who was more interested in theology. He once confronted me with a warning by St. Augustine, who wrote, "A good Christian must beware of mathematicians and those soothsayers who make predictions by unholy methods, especially when their predictions come true, lest they ensnare the soul through association with demons." (The original was, of course, in Latin [1], and I have given a rather free translation; the original dates from around A.D. 400.) The same sentiment was repeated, in a somewhat different form, in Augustine's *Confessions* [2], and it still gets quoted from time to time. (See, for example, [3], p. 167.) If you happen to come across it, you should be aware that, in Augustine's day, "mathematicians" meant what we now call "astrologers." [3]. The old usage seems to have occurred occasionally as recently as the 1700's, although the modern meaning goes back to around 1400.

Notes

[1] *Quapropter bono christiano sive mathematici sive quilibet inpie divinantium, maxime dicentes vera, cavendi sunt, ne consortio daemoniorum animam deceptam pacto quodam societas inretiant.* (*De genesi ad litteram*, Book II, chap. xvii, in vol. 3, part 1 of Augustine's works, edited by J. Zycha, Prague, Vienna, and Leipzig, 1894, pp. 61–62.

[2] Book 4, chap. 1

[3] M. Greenberg, *Euclidean and Non-Euclidean Geometries*, 2nd ed., W. H. Freeman & Co., 1980.

[4] Compare the modern slang use of "mathematician" to mean "card sharp."

Pi Mu Epsilon J.* **9 (1991), 309. Reprinted by permission.

THE TEMPTATION OF PROFESSOR McSHOHAT

According to Hindu tradition, when a sage becomes too saintly the god Indra becomes worried that such a man will end up as a rival god, so he sends a beautiful woman to tempt him. Hence if you hope that your daughter will become extraordinarily beautiful, you may name her Indrasena—Indra's army. Some sages are not attracted to beautiful women, but Indra is resourceful. Consider the case of Professor McShohat.

J. Phipps McShohat was a metallurgist by training, but by means of hard work and an exceptionally vivid imagination he had become an authority on extraterrestrial biology. You may well wonder how anyone can be an authority on extraterrestrial biology, when we have as yet visited only one extraterrestrial body, and that an extraordinarily lifeless one. If so, you underestimate the resourcefulness of the academic mind. Professor McShohat was nothing so crude as an observer of extraterrestrial life; he was a theorist, and a theorist of unusual power. He was admitted to have the most convincing theories of how, when, and why extraterrestrial life would originate, how it would look, and how to communicate with it. It was generally felt that when (not if) such life would be found, it would certainly conform to McShohat's specifications. His logic was impeccable, his mathematical analyses were unchallenged, and his scientific papers in refereed journals numbered in the hundreds. If extraterrestrial life had been discovered, there was no doubt but that a Nobel Prize in Astronomy would have been created to honor him. As it was, he headed the Institute for Exobiology, with a budget of impressive size, which was the next best thing.

It is not surprising that a man of such achievements had little time, and less inclination, to be distracted by beautiful women. Appropriately enough, he was tempted by an Unidentified Flying Object. Of course he kept an eye on reports of UFO's, but there were powerful reasons for not taking them seriously: he could only too easily devise commonplace explanations, sometimes several for each one, for the reported UFO's. A considerable part of his list of publications was devoted to debunking such sightings; they gave him a reputation for hard-headedness that reinforced his theoretical speculations. However, he had interests that did not appear in his public lectures or his published papers. One of these interests was the design of a practical ferry for bringing interplanetary visitors out of orbit. The ideal design, he found, was in the shape of an inverted saucer, with glowing ports around the edge and an adverse effect on internal combustion engines; but he had not published his conclusions since they had vanished into the classified files of the Air Force.

He was, accordingly, somewhat intrigued to receive a letter from someone who signed himself, as nearly as McShohat could decipher his handwriting, Indrasena Nagarakshasha, describing a sighting that exactly corresponded to McShohat's hypothetical vehicle, and mentioning details that I have not repeated here (since, as I said, the whole matter was given a high security classification by the Air Force). The letter had a United Nations stamp, and indeed it appeared that I. N. was attached to the Indian delegation in, apparently, some sort of theological capacity. McShohat was, of course, perfectly able to devise a commonplace explanation of this, like all other sightings. He was well aware, for example, of what can be accomplished with a garbage can lid, some luminous paint, and a good telephoto lens. However, the extraordinary coincidences of detail, and the fact that he had never had a chance to investigate a sighting in India, led him to seek an interview. If he had been a Sanskritist he might have become suspicious, but that he was not.

The writer of the letter turned out to be a petite young woman with a South Indian accent, but wearing a conventional dress, who invited McShohat to call her Indrasena. She seemed to be more interested in discussing the political scene in Andhra Pradesh than in Unidentified Flying Objects, but she did have some interesting eye-witness reports, as well as an unusual lack of theories about them.

It did not take much encouragement to lead Professor McShohat to set out for India. Indrasena, now dressed in a sari, was extremely helpful—she seemed to have the instincts, and the training of a devoted travel agent. She booked McShohat into New Delhi on Air India, wangled him an extra baggage allowance for his investigatory gear, and then escorted him by a succession of increasingly primitive vehicles to a tiny village in the foothills of the Himalayas, a village that was not even marked on the large-scale map that Indrasena had helpfully provided. Here was where the supposed UFO sighting had been made. McShohat's first act was to establish its precise latitude and longitude with the use of the sextant and chronometer that he always carried on such expeditions, and he inserted its location and an approximation to its unpronounceable name on his map. Then he sought confirmation of the reports he had heard. This was readily available, if he could trust an interview with the local blacksmith, an interview that had to be conducted through interpretation by Indrasena of an interpreter of an interpreter. McShohat taped the interview, although in view of the obscurity of the language it is hard to imagine for what purpose.

The smith told a circumstantial story about a chariot that had descended from Heaven on a pillar of fire. Its crew had traded metal artifacts for local goods. The professor could not get a clear description of the crew, and of course there had been no cameras. Unfortunately it appeared that metal was scarce in the village, and the artifacts had all been beaten into plowshares and the like, but with some persuasion and a considerable amount of hard cash, the smith was induced to part with a piece of steel, said to have come from the visitors. Professor McShohat at once attacked this with his portable nuclear magnetic resonance apparatus. It proved to have a composition unlike any steel he had ever heard of, and he had been a recognized authority on the steels of the world.

According to the helpful smith, the chariot of the gods had taken off in a south-westerly direction, so southwest was where McShohat wanted to go. Resourceful as usual, Indrasena arranged for a bullock cart and a relay of drivers, and after a week of extremely uncomfortable travel they arrived at another, equally obscure, village, with another, equally unintelligible, language, where a similar story was repeated. Professor McShohat had occupied himself during the intervening week with devising explanations of what he had heard in the first village, explanations that began to seem less convincing as the next interview added more detail. Here he obtained another piece of metal, which matched the first one in analysis and appearance. McShohat entered the location on his map and then set out again.

So it went for week after week. A week in a bullock cart doesn't get you very far, but the weeks add up, and they covered a substantial distance. The events of the journey were much the same at each stage. Between villages, about a week apart, nobody knew anything about chariots from Heaven, but at the end of the week there was always a village with the same sort of story and the same kind of souvenir from the heavenly visit. Details accumulated: the heavenly visitors had apparently been small of stature, with extra arms, but otherwise humanoid, and there was some indication of a greenish cast to their skins. The only missing detail was the adverse effect on internal combustion engines, but the nearest internal combustion engine was hundreds of miles away.

As the expedition proceeded, the weather become hotter and hotter, and Professor McShohat became increasingly uncomfortable in the business suit that Indrasena insisted that he had to wear in order to impress the villagers with his academic credentials. She herself adapted her costume to local custom, again because (she insisted) it was necessary for her to conform to local manners in order to negotiate successfully for bullock carts, interpreters, and so on. She started in the cool foothills in embroidered jacket and trousers, later returned to saris, and by the time they arrived in southern India she was bare to the waist, in accordance with the old-fashioned manners of the district. Here the chariot of the gods could no longer be tracked because it had disappeared over the Indian Ocean. Indrasena guided the professor to a town in which an automobile was available to take him to where there were airplanes, and left him there; she was, she said, going to visit relatives, and would see him again in the United States.

Professor McShohat returned to the Institute with suitcases full of pieces of metal, tapes, and notebooks, and a very troubled mind. As he attempted to make some sense out of his experiences in India, he had a hard time explaining it all away. Finally he began to suspect that he had something. The least implausible explanation that he could devise was that an extraterrestrial vehicle with a humanoid, but definitely not human, crew, had landed in the North and had proceeded to make exploratory hops of nearly uniform length, each time making gifts to the natives in return for samples of local products. Even with the resources of an excellent laboratory and an extensive library, there was no explanation of that steel except that it had originated in a different world. It seemed hardly possible that those isolated villages could have concocted such consistent stories when intermediate places along the route knew of nothing of the sort.

Eventually Professor McShohat became so convinced that he published a carefully documented paper in the *Journal of Interplanetary Studies*, analyzing what he had found and concluding that the Earth had indeed received alien visitors. The article caused an immediate sensation. The professor was invited everywhere to lecture; he was interviewed on television by Barbara Walters, and appeared with Carl Sagan on the Johnny Carson show.

Then came the crash. The next issue of the *JIS* contained a devastating paper by Professor Vijendra from the Tata Institute which showed that the steel that had so impressed McShohat was indistinguishable from that produced by an obscure subcaste of Indian metal-workers, the Indracharis, who believed that their craft had been revealed to their ancestors by Indra himself. Furthermore, they had migrated from an original home in North India and had established themselves, for religious reasons, at intervals of seven days' travel, indeed precisely where McShohat's meticulously maintained map of his travels had located his finds. Finally, the Indracharis used hallucinatory drugs derived from their sacred plant, and fervently believed that Indra visited them at intervals in a chariot from Heaven, in order to verify that they were following his instructions. Professor Vijendra's credentials were impeccable; the reference material about the Indracharis was verifiable beyond a doubt once someone could be found who could read it; and Professor McShohat was ridiculed everywhere, from the editorial pages of *Nature* to the news conference held by the President of the United States. To add to McShohat's discomfiture, Indrasena had not returned, and nobody at the U.N. would admit to ever having heard of her.

Professor McShohat was forced to resign from the Institute, and was forgotten even faster than he had come to prominence. So it came about that when, as we all know, an alien space-ship actually landed in East Brewster, Massachusetts, he was not there to open conversations with its occupants. Nobody even thought to interview him, except an enterprising reporter from the *Barnstable County Oracle*, who finally tracked him down at a mushroom farm in South Dartmouth, where he was supporting himself as a night watchman. "I don't believe it," was all he would say. When the reporter turned in her story, the editor killed it.

SLEEPER

At most universities nobody except deans, and people who hope someday to become deans, goes to faculty meetings. At the university where I teach, everybody who can possibly do so goes to as many faculty meetings as possible. This is not from a sense of obligation, peculiar to this university, but simply because the average faculty meeting here is the best show around. The show is put on by our two most famous professors: Perkins, the brilliant young psychologist, and Gaylord, the distinguished elderly historian. They hate each other: nothing very uncommon in academic circles. Faculty meetings give them a chance to oppose each other in public, over every issue, however trivial; and both are masters of biting academic invective. This is an art form that you have to hear to appreciate. It loses its quality if you try to transfer it to paper. So we attend faculty meetings, and we are rarely disappointed. The curious thing is that almost nobody in the audience knows why Perkins and Gaylord hate each other. Perkins himself almost certainly doesn't know, except for the reason that Gaylord hates *him*; Gaylord has probably long since forgotten what started it. Perhaps the only one who really remembers is Annette Baldry, who was there when it all began. This is the story that she told me after my first faculty meeting, when I asked her whether she could account for the magnificent exchange of insults that I had just witnessed.

Mithridates Perkins, as a teenager, attended a progressive school that encouraged independent projects by the students. Perkins undertook to do a project on **sleep**. It was his unusual first name that set the direction of his research: he had read up on Mithridates, the king who immunized himself against poisons by taking gradually increasing doses of all of them. Perkins undertook to apply the same principle to sleep. He frequently felt sleepy after lunch, but it is not easy to take a satisfactory nap at a really progressive school. Perkins determined to make himself wake-proof from one to two o'clock every afternoon. He began by having his roommate whisper to him after he had fallen asleep. When he had learned to tolerate whispers, he proceeded to louder and louder noises. After the loudest alarm clock ceased to affect him, he progressed to fire alarms and the tolling of the chapel bell. At this period it was not uncommon for the principal's luncheon guests to be startled by a series of firecracker reports, or a hundred-decibel rendition of the Anvil Chorus. If they asked about the noise, the principal would tell them, soothingly, "Oh, that's just Perkins's nap." It was a very progressive school. At the end of the training period Perkins slept from one to two under all circumstances, and there was a special provision in the fire drill for rescuing him in case of a fire during his nap period. He had published an article about his project in the *Journal für angewandte Psychologie*, and had an appreciative letter from B. F. Skinner in a frame over his desk. He then went to the University. The experiment was successfully over; he wanted to go on to other things; but he still had to have his nap after lunch.

At first all went well; Perkins was careful not to enroll in any one o'clock classes. But like every freshman, he had to take History 31—Gaylord's class—and at the end of the first semester it turned out that the final examination for History 31 was from 1 to 4 P.M.

Gaylord was noted for his terrible temper even then, so Perkins braced himself and went to Gaylord's office. Having been to a progressive school, he naturally began by asking for permission to take the examination at some other time. This was contrary to several centuries of tradition, and Gaylord was firm in his refusal. Perkins then asked for permission to come to the examination an hour late. He was confident, he said, of being able to handle the examination in two hours instead of three. Gaylord was insulted, since nobody was supposed to be able to do his examinations in even three hours. He explained to Perkins that freshmen must not presume to insult their betters, and that he didn't believe Perkins's story anyway. Finally he helpfully suggested No-Doz pills before the examination. Perkins, who of course was conditioned against caffeine as against everything else, saw that he would have to handle the problem himself. He went out and bought a small pillow.

Exam day arrived, one o'clock arrived, and Perkins was there. As the examination papers were being handed out, Perkins put his pillow on his desk, put his head on the pillow and was instantly asleep. And there he was, snoring gently, when Gaylord arrived at a quarter of two to see how the examination was going. Gaylord was deeply incensed. That a freshman should presume to sleep during his examination was an unforgiveable insult, worse than reading newspapers during lectures. He poked Perkins. Nothing happened. He shook him. Nothing happened. He jerked the pillow from under Perkins's head. The head thumped on the desk, but Perkins slept on. It was clear to Gaylord that he was being teased by this student. He swore loudly, and heads came up all around the room. Proctors rushed up to see what was the matter; and Perkins slept on. Gaylord loudly demanded campus police, municipal police, state police, deans, paramedics, and psychiatrists. Proctors rushed off to telephones. People did not trifle with Gaylord, and so it happened that as the clock struck two, Perkins was surrounded by policemen, deans, and hospital attendants with stretchers, all arguing loudly about what should be done, while the class gleefully neglected their examinations. At that moment, of course, Perkins awoke. He picked up his pen and started work on his examination. Gaylord had a heart attack and was carried out on one of the stretchers.

"So I suppose Perkins flunked?" I asked Annette Baldry.

"No," she replied. "Gaylord was in the hospital for two weeks, and I had to grade the examinations. Perkins really was able to handle the exam in two hours. He got an A. That, of course, was the final unforgiveable insult."

A Translation of Mayakovsky's "Hymn to Learning"

Beings of all the kinds collect:
People and birds and centipedes,
With feathers fluffed and spines erect
They gather to watch each others' deeds.

They watch the Sun, observe the seasons,
Survey the sweep's black occupation.
It's all exciting, for no good reason
Except to build their education.

A lecture: that's no human creature,
But merely some bipedal donkey
Describing each minutest feature
Of the Brazilian spider monkey.

"Devour a book;" that, I suppose,
Is just a harmless metaphor,
But I imagine a hapless rose
Crushed in the jaws of an ichthyosaur.

Our backs may twist, absurdly fold us,
But let's not fret about our shape.
Rather, recall what Darwin told us:
We're all descended from an ape.

The heart of a girl, that we cannot keep,
A fragile relic of beauty past,
Is no more reason for us to weep
Than the tail of a comet that doesn't last.

After the night, the Sun will still
Rise and smile at our human rules;
Down on the sidewalk the students mill
Making their way to the science schools.

If measles strikes, and lays you low,
Don't lie there like the stupid brutes.
You've got an education, so
Sit up, and calculate square roots.

Camp Butner (1940)

Last spring in Veazey Ridge and Stem,
In Knap of Reeds and Surl,
About the red tobacco fields
The dogwood cast its pearl.

This year no petals fall upon
Farms that are lying fallow;
Among the muddy weeds and stumps
The tanks and field guns wallow.

Poets

I don't understand poets.
Poets think machines are unromantic because machines are predictable, and give the same result every time.

Poets like people, because people are unpredictable and what they do is therefore interesting.

Poets dislike modern physics because it is mysterious and unpredictable.
They don't like the idea that an electron can get from here to there without our being able to say which way it went.
Poets dislike the idea that an atom can decay at random.

Why don't poets think physics is romantic?

SECTION 16

REVIEWS AND MISCELLANEOUS ARTICLES

ARE MATHEMATICIANS UNNECESSARY?*

Are mathematicians unnecessary? At least, is mathematics no longer of any use to scientists? Can the computer now solve all scientific problems? Is it now a waste of time for students of mathematics to do homework? These and similar questions were debated in a recent series of four articles [1] in the Soviet weekly, **Literaturnaya Gazeta**, a publication not much read by Western mathematicians (including me—my attention was called to the first two articles by [2]). I understand that, as the official organ of the Writers' Union, **Literaturnaya Gazeta** is a very authoritative publication indeed. Its presentation of opposing views is, according to experts on the Soviet Union, to be taken as evidence of a power struggle among very highly-placed people about the role of mathematics and of computers in the Soviet educational system. The eventual outcome promises to be significant for the future of Soviet science. I shall therefore try to summarize the main points here.

Kitaigorodskii, the author of the opening article, "A Divorce case," was originally a physicist, and is now a well-known figure in the fields of history, philosophy and education in science. The "divorce" is the divorce of mathematics from science and technology with the computer as third party. Kitaigorodskii takes the point of view, not uncommon among physicists, that proofs of intuitively plausible results are a waste of time. For example, he pokes fun at mathematicians who study the packing of spheres, since all one has to do is to pack some and look at them. [He does not address the question of how one does this in more than three dimensions.] He claims that all problems of interest in applications can now be solved by computers. Consequently "applied mathematics" is redundant. As for pure mathematics, it is

---

*Math. Intelligencer **2**(4) (1979), 172–173.

now reduced to an art form. Kitaigorodskii does not object to art, but says that only the most capable artists should be encouraged to practice it. [The inherent dangers of this point of view are perhaps more obvious in the West than in the Soviet setting.] Because of the computer, the teaching of mathematics must change—it must be taken away from the mathematicians; moreover, students should never be made to solve any problems since the pocket calculator, or anyway the computer, will provide all the answers.

It is interesting that a rather similar view of the future of mathematics was proposed almost simultaneously in the West by Frauenthal [3]. However, Frauenthal seems to see the computer as forcing mathematics to follow different directions, whereas Kitaigorodskii sees no future at all for mathematics in connection with science.

The second article was by M. Evgrafov, a well-known analyst. This is a rejoinder, although it appears superficially to be an attack on mathematics from a different direction. Evgrafov proposes three "incontrovertible" theses: that no pure mathematics finds any application for 50 years, that 99% of all mathematics is forgotten in 50 years, and that at most 100 people appreciate any given piece of pure mathematics. [He is evidently exaggerating for effect: any mathematical reader can produce counterexamples.] He gives no evidence to support his theses, but he draws some conclusions. A subject with such unpromising characteristics must be good for something, or it could not have survived for thousands of years. Indeed, the best mathematics is very important. However, it is difficult for practical people to take advantage of what mathematics can do for them, because of the difficulty of communicating with mathematicians. Evgrafov suggests the creation of a class of intermediaries who can bridge the gap. In a good Soviet spirit, he wants a plan for making the best use of mathematicians—a difficult task. This leads him to attack the "publish or perish" principle, which is apparently as popular in the Soviet Union as it is in the West. He sees it as a regressive policy that prevents "middle-level" mathematicians from broadening their education while forcing them to attempt to do "creative" work instead.

Evgrafov then points out that, contrary to Kitaigorodskii's claim, computers do not solve problems at all; they just do very rapid calculations, and mathematicians at all levels are needed to put problems into forms that computers can handle. For the most difficult problems this activity requires very highly trained mathematicians, whom Soviet society therefore needs [as, presumably, do other societies.]

In the third article, M. Postnikov, a prominent topologist, deplores the public's ignorance of the accomplishments of mathematics, an ignorance which he attributes to the 17th century level of present-day teaching of mathematics. He says he agrees with Evgrafov (who has, however, overstated his three theses), but Postnikov wants to explain more fully what he thinks mathematics really is. Basically, he says, it consists of schematic models, schematic models of schematic models, and so on—

an idea that he develops at considerable length. Mathematics does not consist of solving special problems for their own sake, but of creating models by the use of which problems can be solved. The effect of the computer has merely been to change the directions and the rate of mathematical progress, something that has always been happening anyway.

So much for Kitaigorodskii's metaphor. As for his specific claims, most of them are simply wrong. Problems on sphere packing are difficult, significant, and important in applications. As Evgrafov has already said, no computer can solve even the simplest problem by itself. Postnikov, however, feels that the computer has had little real effect on the relationship between mathematics and science. It has of course accelerated the obsolescence of old mathematics and the creation of new mathematics, but this is a process that goes on anyway.

Postnikov closes with some inspirational remarks about the possible mathematization of the humanities.

In the final article, four well-known applied mathematicians deplore Kitaigorodskii's position. As people who spend their lives helping practical people solve problems with the help of computers, they do not want to see any "divorce." Abstract ideas are important for the development of science and technology. For example, without complex numbers, Riemannian geometry, group theory, operator theory, etc., there would be no modern physics with all its technological consequences. They point out that there are many new and effective mathematical methods for solving practical problems, but these are methods that can be implemented only on a computer. However, to identify applied mathematics with computing would be like identifying physics with instrumentation.

On the other hand, Kitaigorodskii is right that the computer has changed the way we look at the interaction between mathematics and applications, and the teaching of mathematics should change accordingly; but a sound knowledge of mathematics is essential even if you have a computer in your pocket. The most advanced computer can produce routine solutions only of routine problems.

As for Evgrafov's three theses, the first is totally wrong (it is easy to think of applications now being made of mathematics created since 1930). His second thesis is overstated, but even if it isn't, does it matter? As to the third, one can make similar claims in all fields.

Finally, whereas Postnikov says that the computer has nothing to do with the marital problems of mathematics and science, they say that in fact the computer strengthens the bond between the spouses, and this is all to the good; they call for closer association between the two types of mathematics.

The one point on which all participants seem to agree is that, especially in view of the existence of computers (pocket or otherwise), changes are needed in the teaching of mathematics; but there isn't much agreement, and in fact not much specific suggestion, of what the changes ought to be. Evgrafov, at least, is unimpressed by some proposals for change: he compares the situation to that of a man who still feels hungry after a heavy meal, but is satisfied after dessert, and therefore concludes that meat and vegetables are a waste of time: one should just eat dessert. Although none of the articles says so explicitly, it seems that we need much more experience with varieties of instruction before we can know what would work best under the new circumstances in which we find ourselves.

References

[1] Aleksandr Kitaigorodskiĭ, Delo o razvode (A divorce case), Literaturnaya Gazeta 1979, no. 43 (24 October), p. 13; M. Evgrafov, A byl li brak? (But was there a marriage?), ibid. no. 49 (5 December), p. 12; M. Postnikov, V plenu šlučaĭnyh metafor (Imprisoned by fortuitous metaphors), ibid. 1980, no. 5 (30 January); E. Ventcel, L. Gurin, A. Myškis, L. Sadovskiĭ, Kompyuter—eto ešče ne vše! (The computer—It's still not everything!), ibid., no. 11 (12 March), p. 11.

[2] Vera Rich, Computers can replace maths for scientists, says Soviet professor, Nature 282 (20–27 December 1979), p. 770.

[3] J. C. Frauenthal, Change in applied mathematics is revolutionary, SIAM News, April, 1980, p. 8.

REVIEW OF *THE WORLD OF MATHEMATICS*, BY JAMES R. NEWMAN.*

While it is not customary to review popular books on mathematics in this Bulletin, this one so far exceeds the norm both in range and in sales that it demands notice. (It is undoubtedly the all-time best seller among mathematics books other than textbooks.) A nonmathematician with an amateur's interest in the subject might well wonder at first why he should buy these volumes rather than one of the more compact (and less expensive) popular books, of which there are many excellent ones that have enjoyed a far smaller sale. However, most short popular books on mathematics cover only a limited selection of topics that are not too technical to discuss superficially and are conceded to possess universal appeal. Most of these topics are included here too, but so is much more, and the reader can make his own choice. The subtitle is in a sense misleading, since the contents are much more literature *about* mathematics than mathematics as such. This is of course inevitable in any popular book. A nonmathematician will not learn much mathematics from these volumes, although he's told a great deal about mathematics and about cognate subjects such as mathematicians, physics, logic, and foreign politics; whether this will help him understand what mathematics is about and what mathematicians do is not for a professional mathematician to say. However, there is also a great deal here of value for the professional mathematician, collected from sources that are not on everyone's bookshelf. Some at least of this material will be helpful to teachers and it would be hard to find any mathematician who will not be entertained by some of it, or who will not find something that is new to him.

The contents are highly varied. Some of the selections are actually from the mathematical literature in the strict sense, some are written specifically for the layman, and some are mathematics only by the editor's fiat. Some are extremely interesting, some are exasperating, and some are downright dull. It would be neither practical nor illuminating to list the contents in detail: the following remarks are indicative rather than exhaustive. There are numerous assorted discussions of the nature of mathematics and mathematical thinking, some old-fashioned and some up-to-date. They illustrate the principle that there are at least as many ways of thinking about mathematics as there are mathematicians. There are many biographical and historical selections. Since the chief interest of mathematicians, outside of mathematics, seems to be the personalities of other mathematicians, there is much here to interest the professionals. There are remarks on the general subject of "numbers," ranging from an enquiry into the question of whether birds can count to Dedekind's own account of irrational numbers. There is fascinating material on applied mathematics (my term, not the editor's), much of which seems to be more applications than mathematics: the discovery of Neptune, the problem of determining longitude, the

*Reprinted from *Bulletin of the American Mathematical Society*, **63** (1957), 154–155, by permission of the American Mathematical Society.

periodic table of the elements, Haldane's famous essay *On being the right size*, Eddington on the constants of nature, Malthus on population. There is an assortment of essays on probability and statistics (it is a pity that room could not have been found for Feller's deflation of the St. Petersburg paradox, when so much of the traditional well-meant nonsense about it is included). There is a lucid exposition of Gödel's theorem by Nagel and Newman, and there is an essay that Lewis Carroll would have enjoyed (written especially for this anthology by Nagel) on *Symbolic notation, Haddocks' Eyes, and the dog-walking ordinance*. There are particularly interesting essays on computing machines by von Neumann, Turing and Shannon. Apparently just to show how far one can attempt to go, there are selections from G.D. Birkhoff's writings on ethics and aesthetics. A real novelty is the inclusion of five selections from mathematics in fiction, ranging from *Gulliver's Travels* to *The New Yorker*.

All in all, this is an anthology with the faults of its genre and more virtues than most specimens of its kind, especially in the set of mathematical anthologies of which it is almost the only example. It has as legitimate a place in any mathematician's library as the *Oxford Book of English Verse* has in that of a specialist in English literature.

REVIEW OF *THE MATHEMATICS OF PHYSICS AND CHEMISTRY*, BY HENRY MARGENAU AND GEORGE MURPHY.*

Anyone who has read stories about the South Seas is aware that there is a language called Pidgin English (actually there are several kinds) which seems at first sight to be a clumsy and inept parody of English. It has, as a matter of fact, attained wide currency in some places, and is now recognized as being a genuine language in its own right, although somewhat limited in its vocabulary. Its repulsive aspects (to a user of standard English) are mitigated, if not altogether removed, by this recognition; and it now appears that since Pidgin English serves a useful purpose there is no need to try to replace it by standard English, and indeed little point in attempting to do so.

I have often been irritated, like other professional mathematicians, by the clumsiness and departures from the accepted norm of the mathematics used by physicists and presented in books such as this one. From investigation of this and other books on physics, as well as on the mathematics of physics, I conclude, however, that physicists deal with mathematics of an entirely different kind from that used by mathematicians. In fact, it is not unfair to say that they use pidgin mathematics. They use mathematical reasoning mainly as a crutch, either to convince themselves that they are reasoning correctly in complicated situations or to help themselves remember the sequence of steps in an argument. In either case mathematical rigor is irrelevant, and any kind of plausible argument, however dubious logically, will serve. Since the physicists' conclusions are usually known to be true, and will be discarded if they do not check with experiment, little harm is done. A consequence of the physicists' approach is that there are several kinds of pidgin mathematics, with "theorems" that appear discordant to the outsider. Thus in thermodynamics no multiply-connected regions occur, and the student learns (as in this book) that the integral of an exact differential around a closed path is zero. In hydrodynamics (which is not covered in this book) the situation is different, since multiply-connected regions are common. That pidgin mathematics may lead to mistakes in complicated situations is no reason for deprecating it. Mistakes made by physicists are quickly recognized by their inconsistency with experiment, and a new branch of pidgin mathematics arises to deal with the topics where the old branch fails. The situation differs from the linguistic one in that pidgin mathematics generally seems to have come first, instead of developing out of standard mathematics. It is nevertheless true that pidgin mathematics is a different subject, to be assessed by different standards.

I have, on occasion, argued that physicists should know (not necessarily prove) correct statements of theorems and therefore ought not to say, for example, that (as this book implies) all continuous functions have convergent Fourier series. This

*Reprinted from *Bulletin of the American Mathematical Society*, **65** (1959), 249–251, by permission of the American Mathematical Society.

now seems unreasonable; the mathematics of physics is not expected to be logically valid. Furthermore, the mathematics of one branch of physics is not expected to be valid in a different branch. This perhaps accounts for the authors' remark that Hankel functions are "of interest only in connection with non-integral n," thus (since Y_n is not introduced) leaving one helpless in problems about annuli. Again, the authors' treatment of the calculus of variations is in the metaphysical style of the seventeenth century, with "variations" that are sometimes zero and sometimes not. This is inexcusable if the idea is to prove anything, but perfectly all right if the idea is simply to furnish a mnemonic for Euler's equation.

The authors say in their preface, "The degree of rigor to which we have aspired is that customary in careful scientific demonstrations, not the lofty heights accessible to the pure mathematician." That is, this is a book about pidgin mathematics; as such, it will not appeal to any standard mathematician who may be looking for a text in applications of mathematics. It may, indeed, give him qualms about the validity of "careful scientific demonstrations." I need say little more, since practitioners of pidgin mathematics are unlikely to read this *Bulletin*. The first edition (reviewed in Bull. Amer. Math. Soc. vol. 57 (1945) pp. 508–509) was enormously successful. The second edition differs from the first chiefly by the addition of a section on Fourier and Laplace transforms. Parts of the book are primarily physical (thermodynamics, mechanics of molecules, quantum mechanics, statistical mechanics), some are handbook-style collections of facts (vectors, tensors, coordinate systems, matrices, numerical methods, and the parts of group theory that are too advanced for elementary texts and too special for advanced ones); some consider mathematical tools (differential equations, special functions, calculus of variations, integral equations). The physical parts seem lucidly written and can even be read by mathematicians who want to acquire a smattering of physics to impress their friends. The rest seems adequate within the setting for which it was designed, although even so some physicists have not found it altogether satisfactory as a text; perhaps this corresponds to the fact that (for example) a text written in Melanesian Pidgin would cause difficulties for a reader of Australian Pidgin.

REVIEW OF *PRINTING OF MATHEMATICS/AIDS FOR AUTHORS AND EDITORS AND RULES FOR COMPOSITORS AND READERS AT THE UNIVERSITY PRESS, OXFORD*, BY T.W. CHAUNDY ET AL. AND *MATHEMATICS IN TYPE*, FROM THE WILLIAM BYRD PRESS.*

Printing is a necessary evil; there is substantial agreement among mathematicians that an alleged piece of mathematics has no standing until it has appeared in print for all interested people to read. There is also a general impression that editors make arbitrary and unreasonable rules about the form of manuscripts; and that printers impose absurd restrictions on the symbolism which may be used. These two books are intended to dispel these impressions and give helpful advice to authors. Both of them give detailed explanations of how mathematics actually gets into type; understanding this, the mathematical author can understand why one of two equivalent notations is more economical than another, and generally what he should do (or not do) in order to help the printer. One book or the other should be required reading for anyone who is going to write a mathematical paper. The Oxford one is considerably more detailed about the mechanics of monotype composition and proof correction; in some other respects it may be misleading to an American reader, since of course it represents Oxford practice, which is not entirely typical even of British practice. The Byrd guide is somewhat safer for an American reader to follow, but part of its discussion is based on methods used at the Byrd Press, and not in general use, which make it possible to set on a machine many common combinations which ordinarily require hand work. A minor point, frequently overlooked, is clarified by understanding the mechanics of printing: manuscripts should not be marked in the same way as proofs, since they are handled by the printer in a different way.

Some of the differences between British and American printing customs are interesting. The Americans insist on typewritten copy; the British are quite happy with legible handwriting, and even prefer it under some circumstances. In Oxford practice displayed formulas are numbered on the right; "A very few distinguished mathematicians have numbered their equations on the left: this is exceptional"—but of course is the American standard. Here the British practice is more economical of space if one is willing to agree that a numbered formula need not be displayed, or conversely that ordinary words are admissible in a display. Oxford prefers $\binom{n}{r}$ to ${}_nC_r$ for reasons of style, but hopefully suggests $(n!r)$ as a distinctive replacement for either. The British prefer $\frac{1}{2}$ for looks; the Americans prefer $1/2$ for legibility. The Oxford book suggests several other notational innovations, for instance $\sqrt{a+b\backslash}$

*Reprinted from *Bulletin of the American Mathematical Society*, **61** (1955), 257–260, by permission of the American Mathematical Society.

for $\sqrt{(a + b)}$ (making the radical sign serve as its own bracket, as | does in an absolute value); $\exp_a x$ for a^x in the case of a complicated x; $y_\sharp$ and $y_\flat$ for max y and min y (here there would be trouble if the y's had subscripts: perhaps $\sharp y$ and $\flat y$ would be preferable, but in any case $\flat$ is hard to *write* distinctively. One may well object to any notation that is not convenient both to write and to print clearly). I add one suggestion of my own: write $¡x + y!$ for $\Gamma(x + y + 1)$. Anyone who is repelled by such innovations may recall that the solidus (/) was an innovation less than a century ago.

Both books pay a deserved amount of attention to matters of style, although here again the Oxford book is fuller. There are two aspects of mathematical style, only one of which has to do with the mechanics of printing. This is the fact that some symbols don't combine happily. For instance, consider a_R, where the subscript is almost as big as the main letter; or j_k, which, especially when the whole thing is a subscript, is much harder for the eye to take in than is k_j. Appearance and ease of reading are improved if some care is taken to adjust the symbolism of the demands of printing.

The other aspect of style is essentially literary, and applies even if the paper is not to be printed from type. Both books stress such frequently overlooked points as that a formula is a phrase or sentence of the same language as the rest of the paper, and should be arranged and punctuated as such. They explain the rules governing spacing of symbols, breaking formulas, etc. The Oxford book discusses in detail the preferred usage of punctuation marks, of "I" versus "we," of "assume," "arbitrary," "only," and so on, and compares the relative merits of alternative ways of saying the same thing. These rules are not arbitrary rules, but a summary of the current usage of writers who write clearly and considerately. One can usually recognize good writing, even if one is not aware of the characteristics which make it so; the authors have isolated some of these characteristics. Some of the spirit of this discussion can be felt from the following quotations.

"A good mathematical presentation is one in which the essential information admits of being 'immediately apprehended'; it should not be sufficient merely to say that it is 'all there' for anyone who has the patience and skill to disengage it."

"Some mathematicians (including the writers) will maintain that symbolism can be overdone; that a remorselessly symbolic mathematics need not be the more intelligible. The passage from mind to mind must be made through the reader's eye, and a microscopic notation, all 'jots and tittles,' indices and subscripts, may be as illegible as a macroscopic exposition relying largely on words and phrases. The ideal lies between these, in which an occasional word punctuates the symbolism and a formula or a little knot of symbols breaks the flow of words."

"Mathematicians who are writing in English are asked not to forget the dignity and traditions of the language. What they write purports to be English prose, even though symbols have replaced many of its words; it should be both readable and speakable as well as printable. Thus symbols such as '∴,' '∵,' or end-tags like 'q.e.d.' 'q.e.f.' are best left behind in the schoolroom. What they say can be as well said in plain English. When 'with respect to' grows tedious by repetition, it need not be cut to 'w.r.t.,' which is not current English. The preposition 'in' will generally serve."

I recommend this part of the Oxford book especially to all of us who feel that doing research is so much more fun (and claim that it is so much more important) than writing it down carefully for others to comprehend.

Both books contain long lists of available characters; there is a wealth of choice available for anyone who is imaginative enough to do something besides varying a few letters by covering them with hats of various shapes. The Oxford University Press is, however, not an entirely safe guide for authors since, for example, it is willing to allow, and indeed has allowed, a Chinese character as a mathematical symbol. Other, less well-equipped, presses would disagree; and in general it seems that notations should, if possible, be chosen so that they can be reproduced by all reasonable printers.

REMINISCENCES BY STUDENTS

Professor Boas supervised both informally and formally a number of doctoral dissertations, informally at Harvard as we were informed earlier in the obituary by Philip J. Davis, and formally at Northwestern. The following is the list of Boas' Ph.D. students at Northwestern: José Maria Gonzalez-Fernandez (1958), Daniel Saltz (1959), Masakiti Kinukawa (1960), Alberto Saenger (1962), K. Raman Unni (1963), Christopher Olutunde Imoru (1971), Dale Henry Mugler (1974), Carl L. Prather (1977), Robert Porter (1978), Antoinette Trembinska (1985), and William Butterworth (1987).

We include here some reminiscences from a few of these former students.

REMINISCENCE*

Dale H. Mugler
The University of Akron

For once, my timing was perfect. I had completed the preliminary examinations for the doctoral program in mathematics at Northwestern University in the Fall of 1971. Now it was time to choose a thesis advisor. I can remember standing in one of the large graduate offices in the basement of Lunt Hall and getting some advice from one of the older graduate students. He told me that Professor Boas would soon be stepping down as chair of the department, and might be willing to take on a thesis student. That turned out to be a turning point in my life. I made an appointment to speak with him fairly quickly after that, and when he agreed to be my thesis advisor, what started was a relationship that didn't end when I obtained my degree. My relationship with him continued from that first moment, not only to the completion of my thesis in 1974, but for as long as he was alive.

Not quite ten years ago, I applied for a research fellowship from the Alexander von Humboldt Foundation, whose main support came from the West German government. I received that fellowship and spent an inspirational year in Aachen. My main German contact told me later that one of the key factors that got me that fellowship was Boas's letter of reference. Boas was held in high regard by the German scientists, and they trusted his judgment.

While I was in Germany, I did my best to learn German and converse in the language of the country. The word for thesis advisor in German is different from what we use in the United States, and conveys a different sense of relationship. The word in German is "Doktorvater," which translated literally means *doctor-father* or something like *father of the doctorate*. When I first heard that word used, I thought of how appropriate that was for me.

---

*From the Memorial Service held at Northwestern University

The idea of a father for a doctorate was just about right for how I felt Boas treated me—someone who gave gentle guidance, a person who quickly responded to me when I needed him to explain something, and someone I respected and whose guidance I trusted. It was for me more like the way a father treats his child.

At this stage in my life, I've been on so many tenure and promotion committees, that when I think of what Ralph P. Boas Jr. did for me, it naturally falls into the categories of research and teaching. I'm sure that you can imagine how much Boas helped me in the area of research, for that is the main function of a thesis advisor. Luckily for me, however, our relationship had a lot to do with teaching too.

I'm not sure how it happened, whether it was something that Boas felt was important or whether it was the plan of the professor who assigned teaching assistants. In any case, when I became Boas's thesis student, I also became one of his teaching assistants.

One of my personal memories of serving as a teaching assistant to Boas was the time I looked up during one of my 8 a.m. recitation classes and saw him watching through the door. I was really surprised. He really cared what we did in those classes and he was willing to come watch and see what we did. He had some suggestions for me, too. Looking back, they were suggestions that I have used throughout my own teaching career, and have passed along to some students of my own. In my later years at Northwestern, Boas would call on me occasionally to substitute for him in a lecture when he had to be away. I think I learned how to organize my own lectures when he helped me prepare to teach those classes.

Long after I left Northwestern, Boas still gave me valuable advice on directions for my research and inspired me with his teaching. To be included with this collection of his work is a talk he gave as a major address at the joint MAA-AMS meetings in Phoenix, not long ago. At the end of this talk, he said that he was old and wouldn't do more on those mathematical problems. He was leaving it to the younger people to continue. I like to think that those of us with the good fortune to be his thesis students are continuing what he had in mind.

REMINISCENCE

Carl L. Prather
Virginia Polytechnic Institute and State University

I was at Northwestern University during the time period 1972–1977. The death of Professor Boas saddened me greatly. The mathematical community lost a giant figure who provided valuable services to it for many years and on many levels.

When I initially started working on a thesis with Boas in 1974, he wrote out a list (which I still have) of five possible problems to work on. My failure to work on some of them represented missed opportunities for me, for these problems were subsequently solved and published by others. I found his insistence that I meet with him weekly to discuss mathematical progress beneficial for me. It motivated me to work hard in order to try to have something new to say. Boas had recently edited Volumes I and II of collected papers of George Pólya, which were published in 1974. Boas suggested that I read some of Pólya's papers on uniform approximation of entire functions by polynomials whose zeros are all real or lie on rays. I did so and was led to subsequent work on this topic by J. Korevaar and eventually to a thesis.

I vividly recall Boas on several occasions telling me a story about one of his early experiences as a mathematician. During the late 1930's, he was a visiting scholar at Princeton. While there, he consulted with a well-known mathematician. On one occasion, Boas stated his intention to work on a certain problem, which, as I recall the story, the mathematician said was "impossible." Shortly afterwards, Boas solved the problem and stated his results. The mathematician's response was that the problem was "trivial." At the time that Boas told me the story, I considered it strange behavior for a mathematician. Little did I know....

After finishing my Ph.D. and starting an academic career, I corresponded with Boas often. Over the years, Boas wrote me many letters suggesting problems of interest or bringing my attention to certain papers. On one occasion, I spoke to Boas about some results of N. Obrechkoff, announced in a paper published in French, but whose proofs were given in a long paper written in Bulgarian. To my surprise, Boas told me that he knew Bulgarian, that I should send him the paper with the proofs, and that he would translate the relevant material, which he did. During the 1980's and 1990's, I was amazed to see the number of translations of books and papers from Russian to English that Boas translated which were published.

During the mid-1970's at Northwestern, Boas hired me during part of the summer to take care of his rare and exotic flower collection while he and his wife were out of town. At their home, there was an abundance of such flowers in their joint study, living room, and dining room as well as in the garden. Recalling this, in 1991 I sent them exotic flowers to commemorate their 50th wedding anniversary. They sent me a very kind note subsequently. Unfortunately (for me), this was the last correspondence that I received from them.

I will greatly miss him.

REMINISCENCE

Antoinette M. Trembinska
The City University of New York

"Be wise, generalize!" was the quick-witted advice Ralph Boas offered to me on many an occasion. Although he was my thesis advisor at Northwestern from 1983–1985, I first came to know Professor Boas in the fall of 1980. I was one of his teaching assistants in a calculus course, and I recall that he instructed us to do something we thought unusual. During the exams, students had the option to ask us if their final answer to a problem was correct. So we scrambled around a small classroom in Lunt Hall saying "Yes, it's correct" and the student went confidently on to tackle the next exam problem, or "No, try again" and the student reviewed the work, searching for the error. Professor Boas said that his goal was to improve the students' chances for success on the exam by correcting errors before their work was graded. Whether or not this method was effective, I couldn't say, but what impressed us the most was that despite decades of experience, he sustained a continued interest in incorporating new methods into his classroom teaching. Professor Boas believed that if his students weren't grasping a particular mathematical concept, then he wasn't presenting the material effectively. Professor Boas wrote a short article entitled "Boxing the Chain Rule" (*California MathematiCs* **7**(1) (1982), 36). In it he proposes a new way of teaching students how to differentiate composite functions. It is an effective method that is popular with students, and once you've discovered it, you'll find yourself adapting the method to other problems as well.

As a thesis advisor, Professor Boas was generous with his time and knowledge. I recall that it was Monday when he agreed to oversee my dissertation project, and on Friday of the same week we were already meeting to discuss possible projects. Many months later when I proved my first result, he assigned an exercise for me to complete by our next meeting: to develop five nontrivial extensions of this result. In fact this was the first of many times he offered the sage advice I mentioned earlier,

"Be wise, generalize!" To Professor Boas, finding new problems to work on was an important skill he wanted his graduate students to develop. And he practiced what he preached. If you take a look at his article "A Uniqueness Theorem for Harmonic Functions" (*J. Approx. Theory* **5** (1972), 425–427), you will find the statement "Theorem 1 suggests a number of problems, for example the following." Professor Boas then goes on to enumerate four open problems, one of which he proves, while the remaining three were snatched up by other mathematicians (myself included!).

He was also a very pragmatic person. He strongly recommended publishing your results as quickly as possible in case someone else might be working on the same problem. He emphasized accurate and simple exposition. In the article "How to Publish Mathematics" (*AWM Newsletter* **14** (1984), 9–12) he writes "long convoluted sentences, bifurcating into a plethora of dependent clauses, especially those with verbs deferred to the end, with the consequent effect of demanding close attention from the reader, as well as comprehension of sesquipedalian and abstruse words, or of highly specialized technical jargon, are rebarbative and should be sedulously avoided." Professor Boas wrote this article while I was working on my dissertation, and I recall that he asked me to review his work before he submitted it for publication. I made comments and suggestions about the piece, and he expressed his gratitude. The fact that he had asked me, a graduate student, to critique his work, was enough of a compliment, and I was flattered by his request alone. But then the article appeared in print and you can imagine my surprise and delight when I read "I am grateful to A. M. Trembinska for helpful comments on earlier drafts of this article."

In a similar incident, I relayed to Professor Boas a situation in a calculus class where we considered the concavity of the function $|x - 2|^{1/2}$. A student pointed out that the graph of this function does "hold water" yet the second derivative test proves the function to be concave down on $(-\infty, 2) \cup (2, \infty)$. Professor Boas found this inconsistency interesting and wrote the cleverly entitled article "Does 'holds water' Hold Water?" (*College Math. J.* **17** (1986), 341). Again, at the end of the article he expresses his gratitude to me for having made him aware of the problem.

A few years later I was organizing some papers and created a file folder that contained ideas for possible research projects, all of which were suggested by Professor Boas. I labeled the outside of the folder "Some Ideas," placed it in the file drawer and walked away. A few seconds later I thought of Professor Boas and his commitment to the adage "Give credit where credit is due." So I returned, removed the folder, and added parenthetically to its label "Boas' "!

Some of his articles have become so well-known and liked (see for instance "Snowfalls and Elephants, Pop Bottles and π," in the Northwestern publication *Arts and Sciences* **2**(1) (1979), 2–5), that mere mention of the title brings about a smile or a response such as "Oh, you caught that one too!". When I interviewed for a position at St. Olaf College, my undergraduate talk was based on a Boas article entitled "Travelers' Surprises" (*Two-Year College Math. J.* **10** (1979), 82–88). The article contains a graph with the expression "aha!" pointing to a particular segment of the graph. I began my talk confidently, having reassured myself that I was the only one in the room who knew the material inside and out and was therefore a few comfortable steps ahead of my audience. But at one point in my talk a professor raised his hand and asked "Is this the 'aha!' part of the proof?" and laughter spread throughout the room. Needless to say their familiarity with the article took the wind out of my sails for a few moments, and when I relayed this incident to Professor Boas, he was humbly yet noticeably pleased by their reaction.

During my third year in graduate school I recall asking Professor Boas what his plans were for me in terms of completing my studies. He said that he had in mind that I would graduate the following year. So I asked him what would happen if I wanted to stay on another year. He in turn said that if the university continued to support me, he had no objections, but warned that I wasn't getting any younger! (I was twenty-four years old at the time.)

One winter I attended the joint meetings in Louisville, where I had the opportunity to meet Professor Mary Ellen Rudin at an event sponsored by the AWM. At that time I knew very few female mathematicians, so although I was encouraged by Professor Rudin's warm inviting face, soft gray curly hair, and welcoming disposition, it took me a few moments to reconcile the daunting image I had of her as a mathematician with her physical presence. When I conveyed this to Professor Boas I added warmly that I thought Professor Rudin looked like anyone's grandmother. He paused for a moment, peered over his glasses and said "That's about right, because I think of her as just about anyone's sister." My face reddened while Professor Boas sat quietly, quite content with his response.

Professor Boas was also very supportive of women in mathematics. In fact it was the Boases who introduced me to the AWM and its newsletter. At the winter meeting in San Francisco a few years ago, I organized a special session in entire function theory and when I consulted Professor Boas on the slate of invited speakers, he was the one to suggest another female presenter and was mindful of the gender balance. One summer, while still at Northwestern, I taught algebra and trigonometry to gifted sixth, seventh, and eighth graders. I mailed Professor Boas some of the problems we were working on and apparently one had to do with the setting of a wife meeting her husband, who was coming home from work, at the train station. Professor Boas wrote back "Why didn't you for a change have the wife arriving by train, being met by her husband?" He was exactly right.

I know I speak for several of his students when I say that we were in awe of Ralph Boas, and our admiration is reflected to this day through imitation. We aim to be the effective lecturer that he was, and in our publications we strive to duplicate his casual yet succinct writing style. Although none of us has taken to wearing bow ties, some of us continue to write with green pens, as he always chose to do. If I were to describe Ralph Boas with one word, that word would be "welcoming." He welcomed us into the world of mathematics as eagerly and as warmly as he welcomed us into his office in Lunt Hall. And he extended this welcome to everyone. It is fitting then that he chose to entitle the last book he wrote "Invitation to Complex Analysis."

For those of you fortunate enough to have known Ralph Boas, I hope my reminiscences here have brought to mind a few of your own fond memories. And for those of you who didn't know him, I hope you've caught a glimpse of what you've missed.

RALPH P. BOAS, JR. BIBLIOGRAPHY

[1] A theorem on analytic functions of a real variable, *Bull. Amer. Math. Soc.* **41** (1935), 233–236.

[2] Necessary and sufficient conditions in the moment problem for a finite interval, *Duke Math. J.* **1** (1935), 449–476.

[3] Some theorems on Fourier transforms and conjugate trigonometric integrals, *Trans. Amer. Math. Soc.* **40** (1936), 287–308.

[4] The derivative of a trigonometric integral, *J. London Math. Soc.* **12** (1937), 164–165.

[5] Asymptotic relations for derivatives, *Duke Math. J.* **3** (1937), 637–646.

[6] Some gap theorems for power series, *Duke Math. J.* **4** (1938), 176–188.

[7] Tauberian theorems for $(C, 1)$ summability, *Duke Math. J.* **4** (1938), 227–230.

[8] (H. Pétard, pseud.; Boas with F. Smithies) A contribution to the mathematical theory of big game hunting, *Amer. Math. Monthly* **45** (1938), 446–447; reprinted with additions, *Eureka* (Cambridge, England), 1939.

[9] Representations for entire functions of exponential type, *Ann. of Math.* **39** (1938), 269–286; **40** (1939), 948.

[10] (with S. Bochner) Closure theorems for translations, *Ann. of Math.* **39** (1938), 287–300.

[11] (with F. Smithies) On the characterization of a distribution function by its Fourier transform, *Amer. J. Math.* **60** (1938), 523–531.

[12] (with J. W. Tukey) A note on linear functionals, *Bull. Amer. Math. Soc.* **44** (1938), 523–528; **46** (1940), 566.

[13] A Tauberian theorem connected with the problem of three bodies, *Amer. J. Math.* **61** (1939), 161–164.

[14] (with D. V. Widder) The iterated Stieltjes transform, *Trans. Amer. Math. Soc.* **45** (1939), 1–72.

[15] The closure of translations of almost periodic functions (abstract), *Bull. Amer. Math. Soc.* **45** (1939), 68.

[16] (with S. Bochner) On a theorem of M. Riesz for Fourier series, *J. London Math. Soc.* **14** (1939), 62–73.

[17] Remarks on a theorem of B. Lewitan, *Mat. Sbornik (N.S.)* **5**(47) (1939), 185–188.

[18] The Stieltjes moment problem for functions of bounded variation, *Bull. Amer. Math. Soc.* **45** (1939), 399–404.

[19] Oscillating functions, *Duke Math. J.* **5** (1939), 394–400.

[20] On a generalization of the Stieltjes moment problem, *Trans. Amer. Math. Soc.* **46** (1939), 142–150.

[21] A trigonometric moment problem, *J. London Math. Soc.* **14** (1939), 242–244.

[22] General expansion theorems, *Proc. Nat. Acad. Sci. (U.S.A.)* **26** (1940), 139–143.

[23] (with D. V. Widder) An inversion formula for the Laplace integral, *Duke Math. J.* **6** (1940), 1–26.

[24] Entire functions bounded on a line, *Duke Math. J.* **6** (1940), 148–169; **13** (1946), 483–484.

[25] A completeness theorem, *Amer. J. Math.* **62** (1940), 312–318.

[26] Some uniqueness theorems for entire functions, *Amer. J. Math.* **62** (1940), 319–324.

[27] Some uniformly convex spaces, *Bull. Amer. Math. Soc.* **46** (1940), 304–311.

[28] Univalent derivatives of entire functions, *Duke Math. J.* **6** (1940), 719–721.

[29] Expansions of analytic functions, *Trans. Amer. Math. Soc.* **48** (1940), 467–487.

[30] (with D. V. Widder) Functions with positive differences, *Duke Math. J.* **7** (1940), 496–503.

[31] Functions with positive derivatives, *Duke Math. J.* **8** (1941), 163–172.

[32] A general moment problem, *Amer. J. Math.* **63** (1941), 361–370.

[33] (with G. Pólya) Generalizations of completely convex functions, *Proc. Nat. Acad. Sci. (U.S.A.)* **27** (1941), 323–325.

[34] A note on functions of exponential type, *Bull. Amer. Math. Soc.* **47** (1941), 750–754.

[35] Review of Levinson, Norman, *Gap and density theorems*, in *Bull. Amer. Math. Soc.* **47** (1941), 543–547.

[36] Inversion of a generalized Laplace integral, *Proc. Nat. Acad. Sci. (U.S.A.)* **28** (1942), 21–24.

[37] Generalized Laplace integrals, *Bull. Amer. Math. Soc.* **48** (1942), 286–294.

[38] (with G. Pólya) Influence of the signs of the derivatives of a function on its analytic character, *Duke Math. J.* **9** (1942), 406–424.

[39] Entire functions of exponential type, *Bull. Amer. Math. Soc.* **48** (1942), 839–849.

[40] (with Mary L. Boas and N. Levinson) The growth of solutions of a differential equation, *Duke Math. J.* **9** (1942), 847–853.

[41] (with A. C. Schaeffer) A theorem of Cartwright, *Duke Math. J.* **9** (1942), 879–883.

[42] Representation of functions by Lidstone series, *Duke Math. J.* **10** (1943), 239–245.

[43] Functions of exponential type, I–IV, *Duke Math. J.* **11** (1944), 9–15, 17–22, 507–511, 799.

[44] A differential inequality, *Bull. Amer. Math. Soc.* **51** (1945), 95–96.

[45] (with M. Kac) Inequalities for Fourier transforms of positive functions, *Duke Math. J.* **12** (1945), 189–206.

[46] Functions of exponential type, V, *Duke Math. J.* **12** (1945), 561–567.

[47] (with Ralph P. Boas) Shakespeare's *Twelfth Night*, II, iii, 25–27, *The Explicator*, **3**(4) (1945), no. 29.

[48] Concerning the sequence $\{\cos n_k x\}$, $n_k \to \infty$, *Math. Notae* **5** (1945), 41.

[49] Fundamental sets of entire functions, *Ann. of Math.* **47**(2) (1946), 21–32, **48** (1947), 1095.

[50] (with H. Pollard) Properties equivalent to the completeness of $\{e^{-t}t^{\lambda_n}\}$, *Bull. Amer. Math. Soc.* **52** (1946), 348–351.

[51] A density theorem for power series, *Amer. J. Math.* **68** (1946), 319–320.

[52] The rate of growth of analytic functions, *Proc. Nat. Acad. Sci. (U.S.A.)* **32** (1946), 186–188.

[53] The growth of analytic functions, *Duke Math. J.* **13** (1946), 471–481.

[54] Density theorems for power series and complete sets, *Trans. Amer. Math. Soc.* **61** (1947), 54–68.

[55] Poisson's summation formula in L^2, *J. London Math. Soc.* **21** (1947), 102–105.

[56] (with H. Pollard) Complete sets of Bessel and Legendre functions, *Ann. of Math.* **48**(2) (1947), 366–384.

[57] Sur les suites vérifiant des inégalités portant sur leurs différences, *C. R. Acad. Sci. Paris* **224** (1947), 1683–1685.

[58] Inequalities for the coefficients of trigonometric polynomials, I, II, *Nederl. Akad. Wetensch. Proc.* **50** (1947), 492–495, 759–762.

[59] More inequalities for Fourier transforms, *Duke Math. J.* **15** (1948), 105–109.

[60] (with R. C. Buck and P. Erdős) The set on which an entire function is small, *Amer. J. Math.* **70** (1948), 400–402.

[61] (with H. Pollard) The multiplicative completion of sets of functions, *Bull. Amer. Math. Soc.* **54** (1948), 518–522.

[62] (with K. Chandrasekharan) Derivatives of infinite order, *Bull. Amer. Math. Soc.* **54** (1948), 523–526, 1191; *Proc. Amer. Math. Soc.* **2** (1951), 422.

[63] Quelques généralisations d'un théorème de S. Bernstein sur la dérivée d'un polynôme trigonométrique, *C. R. Acad. Sci. Paris* **227** (1948), 618–619.
[64] Basic sets of polynomials I, *Duke Math. J.* **15** (1948), 717–724.
[65] A class of gap theorems, *Duke Math. J.* **15** (1948), 725–728.
[66] Exponential transforms and Appell polynomials, *Proc. Nat. Acad. Sci. (U.S.A.)* **34** (1948), 481–483.
[67] An upper bound for the Gontcharoff constant, *Duke Math. J.* **15** (1948), 953–954.
[68] Basic sets of polynomials II, *Duke Math. J.* **16** (1949), 145–149.
[69] The completeness of some sets of analytic functions (in Russian), *Izvestiya Akad. Nauk SSSR. Ser. Math.* **13** (1949), 55–60.
[70] Sur les séries et intégrales de Fourier à coefficients positifs, *C. R. Acad. Sci. Paris* **228** (1949), 1837–1838.
[71] Representation of probability distributions by Charlier series, *Ann. Math. Statist.* **20** (1949), 376–392.
[72] The Charlier B series, *Trans. Amer. Math. Soc.* **67** (1949), 206–216.
[73] (with R. P. Agnew) An integral test for convergence, *Amer. Math. Monthly* **56** (1949), 677–678.
[74] Fourier series with a sequence of positive coefficients, *Acta Sci. Math. Szeged, Pars B* (1950), 35–37.
[75] Polynomial expansions of analytic functions, *J. Indian Math. Soc. (N.S.)* **14** (1950), 1–14.
[76] Differential equations of infinite order, *J. Indian Math. Soc. (N.S.)* **14** (1950), 15–20.
[77] Sur une équation fonctionnelle, *Elemente der Math.* **5** (1950), 85–86.
[78] A note on series of positive terms, *J. Univ. Bombay (N.S.)* **19**, part 3, sect. A (1950), 12.
[79] Partial sums of Fourier series, *Proc. Nat. Acad. Sci. (U.S.A.)* **37** (1951), 414–417.
[80] Integrability of trigonometric series I, *Duke Math. J.* **18** (1951), 787–793.
[81] Completeness of sets of translated cosines, *Pacific J. Math.* **1** (1951), 321–328.
[82] Convergence of series and integrals, *Math. Gaz.* **35** (1951), 258–259.
[83] Integrability of trigonometric series II, *Math. Z.* **55** (1952), 183–186.
[84] Growth of analytic functions along a line, *Proc. Nat. Acad. Sci. (U.S.A.)* **38** (1952), 503–504.
[85] Sums representing Fourier transforms, *Proc. Amer. Math. Soc.* **3** (1952), 444–447.
[86] Integrability along a line for a class of entire functions, *Trans. Amer. Math. Soc.* **73** (1952), 191–197.
[87] Integrability of trigonometric series, III, *Quart. J. Math., Oxford Ser.* **3**(2) (1952), 217–221.

[88] Sur les fonctions possèdant une suite de dérivées positives, *Bull. Sci. Math.* **76**(2) (1952), 142–144.

[89] Inequalities between series and integrals involving entire functions, *J. Indian Math. Soc. (N.S.)* **16** (1952), 127–135.

[90] Oscillation of partial sums of Fourier series, *J. Analyse Math.* **2** (1952), 110–125.

[91] Integral functions with negative zeros, *Canadian J. Math.* **5** (1953), 179–184.

[92] Two theorems on integral functions, *J. London Math. Soc.* **28** (1953), 194–196.

[93] Functions which are odd about several points, *Nieuw Arch. Wiskunde* **1**(3) (1953), 27–32; **5** (1957), 25.

[94] Some elementary theorems on entire functions, *Rend. Circ. Mat. Palermo* **1**(2) (1952), 323–331 (issued 1953).

[95] Remarks on a moment problem, *Studia Math.* **13** (1953), 59–61.

[96] Asymptotic properties of functions of exponential type, *Duke Math. J.* **20** (1953), 433–448.

[97] A Tauberian theorem for integral functions, *Proc. Cambridge Philos. Soc.* **49** (1953), 728–730.

[98] *Entire functions*, Academic Press, 1954.

[99] Order of magnitude of Fourier transforms, *Michigan Math. J.* **2** (1953), 141–142 (issued 1955).

[100] Forms for literature citations (letter to the editor), *Science* **120** (1954), 1038.

[101] Moments of analytic functions, *Proc. Amer. Math. Soc.* **6** (1955), 412–413.

[102] Growth of analytic functions along a line, *J. Analyse Math.* **4** (1955), 1–28.

[103] Review of Chaundy, T. W., P. R. Barrett, and Charles Batey, *The printing of mathematics/Aids for authors and editors and rules for compositors and readers at the University Press, Oxford*, in *Bull. Amer. Math. Soc.* **61** (1955), 257–260.

[104] Isomorphism between H^p and L^p, *Amer. J. Math.* **77** (1955), 655–656.

[105] Interference phenomena for entire functions, *Michigan Math. J.* **3** (1955), 123–132.

[106] Review of Hastings, C. Jr., *Approximations for digital computers*, in *Bull. Amer. Math. Soc.* **61** (1955), 462–463.

[107] Absolute convergence and integrability of trigonometric series, *J. Rational Mech. Anal.* **5** (1956), 621–632.

[108] (with R. C. Buck) Polynomials defined by generating relations, *Amer. Math. Monthly* **63** (1956), 626–632.

[109] Inequalities for functions of exponential type, *Math. Scand.* **4** (1956), 29–32.

[110] (with J. M. Gonzalez-Fernandez) Integrability theorems for Laplace-Stieltjes transforms, *J. London Math. Soc.* **32** (1957), 48–53.

[111] Review of Newman, James R., *The world of mathematics*, in *Bull. Amer. Math. Soc.* **63** (1957), 154–155.

[112] "If this be treason...," *Amer. Math. Monthly* **64** (1957), 247–249.

[113] Inequalities for asymmetric entire functions, *Illinois J. Math.* **1** (1957), 94–97.

[114] *Harmonic analysis and entire functions*, Symposium on Harmonic Analysis and Related Integral Transforms, Final Report, Cornell University, 1957, 13 pp.

[115] (with A. C. Schaeffer) Variational methods in entire functions, *Amer. J. Math.* **79** (1957), 857–884.

[116] Letter to the editor, *Scientific American* **197**(1) (1957), 16.

[117] Growth of derivatives of entire functions, *Math. Z.* **68** (1957), 296–298.

[118] Russian translations (letter to the editor), *Science* **125** (1957), 1260–1261.

[119] (A. C. Zitronenbaum, pseud.) Bisecting an area and its boundary, *Math. Gaz.* **33** (1959), 130–131.

[120] On generalized averaging operators, *Canad. J. Math.* **10** (1958), 122–126.

[121] (with A. C. Schaeffer) New inequalities for entire functions, *J. Math. Mech.* **7** (1958), 191–205.

[122] (with R. C. Buck) *Polynomial expansions of analytic functions*, Ergebnisse d. Math. (N.S.), no. 19, Springer-Verlag, 1958, 1964.

[123] Review of Reichmann, W. J., *The fascination of numbers*, in *Amer. Sci.* **46** (1958), 86A.

[124] Almost completely convex functions, *Duke Math. J.* **25** (1958), 193–195.

[125] A variational method for trigonometric polynomials, *Illinois J. Math.* **3** (1959), 1–10.

[126] Representations for completely convex functions, *Amer. J. Math.* **81** (1959), 709–714.

[127] Review of de Bruijn, N. G., *Asymptotic methods in analysis*, in *Bull. Amer. Math. Soc.* **65** (1959), 160–163.

[128] Review of Margenau, Henry, and George Moseley Murphy, *The Mathematics of Physics and Chemistry*, 2nd ed., in *Bull. Amer. Math. Soc.* **65** (1959), 249–251.

[129] Beurling's test for absolute convergence of Fourier series, *Bull. Amer. Math. Soc.* **66** (1960), 24–27.

[130] *A primer of real functions*, Carus Mathematical Monographs, no. 13, Math. Assoc. of America, 1960, 1981.

[131] On some versions of Taylor's theorem, *Enseignement Math.* **5**(2) (1959), 246–248 (issued 1960).

[132] Inequalities for monotonic series, *J. Math. Anal. Appl.* **1** (1960), 121–126.

[133] A series considered by Ramanujan, *Arch. Math.* **11** (1960), 350–351.

[134] Differentiability of jump functions, *Colloq. Math.* **8** (1961), 81–82.

[135] (with S. Izumi) Absolute convergence of some trigonometric series I, *J. Indian Math. Soc.* **24** (1960), 191–210.

[136] An inequality for nonnegative entire functions, *Proc. Amer. Math. Soc.* **13** (1962), 666–667.
[137] (with Q. I. Rahman) Some inequalities for polynomials and entire functions (in Russian), *Dokl. Akad. Nauk SSSR* **147** (1962), 11–12.
[138] (with Q. I. Rahman) L^P inequalities for polynomials and entire functions, *Arch. Rational Mech. Anal.* **11** (1962), 34–39.
[139] Inequalities for polynomials with a prescribed zero, *Studies in Mathematical Analysis and Related Topics, Essays in Honor of George Pólya*, Stanford University Press, 1962, pp. 42–47.
[140] Inversion of Fourier and Laplace transforms, *Amer. Math. Monthly* **69** (1962), 955–960.
[141] Integrability of nonnegative trigonometric series, *Tôhoku Math. J.* **14** (1962), 363–368.
[142] Majorant problems for trigonometric series, *J. Analyse Math.* **10** (1962), 253–271.
[143] On sine series with positive coefficients, *Math. Z.* **80** (1962), 382–383.
[144] (with Q. I. Rahman) Inequalities for monotonic entire functions, *Michigan Math. J.* **10** (1963), 225–230.
[145] Yet another proof of the fundamental theorem of algebra, *Amer. Math. Monthly* **71** (1964), 180.
[146] Indefinite integration by residues, *Amer. Math. Monthly* **71** (1964), 298–300, 906.
[147] (with P. M. Anselone) The atypical zeros of a class of entire functions, *J. Math. Anal. Appl.* **8** (1964), 278–281.
[148] The distance set of the Cantor set, *Bull. Calcutta Math. Soc.* **54** (1962), 103–104.
[149] Periodic entire functions, *Amer. Math. Monthly* **71** (1964), 782.
[150] (with V. C. Klema) A constant in the theory of trigonometric series, *Math. Comp.* **18** (1964), 674.
[151] (Portions of) A Symposium on the publication of mathematical literature, *SIAM Rev.* **6** (1964), 431–454.
[152] Integrability of nonnegative trigonometric series II, *Tôhoku Math. J.* **16** (1964), 368–373.
[153] Tannery's theorem, *Math. Mag.* **38** (1965), 66.
[154] More about quotients of monotone functions, *Amer. Math. Monthly* **72** (1965), 59–60.
[155] Quasi-positive sequences and trigonometric series, *Proc. London Math. Soc.* **14A**(3) (1965), 38–46.
[156] (with L. Schoenfeld) Indefinite integration by residues, *SIAM Rev.* **8** (1966), 173–183.
[157] Fourier series with positive coefficients, *Bull. Amer. Math. Soc.* **72** (1966), 863–865.

[158] (with L. Schoenfeld) Indefinite integration by residues, II, *Amer. Math. Monthly* **73** (1966), 881.

[159] (H. Pétard, pseud.) A brief dictionary of phrases used in mathematical writing, *Amer. Math. Monthly* **73** (1966), 196–197.

[160] Lipschitz behavior and integrability of characteristic functions, *Ann. Math. Statist.* **38** (1967), 32–36.

[161] Fourier series with positive coefficients, *J. Math. Anal. Appl.* **17** (1967), 463–483.

[162] (with R. Askey) Fourier coefficients of positive functions, *Math. Z.* **100** (1967), 373–379.

[163] Note on integration by residues, *Elemente der Math.* **22** (1967), 106.

[164] Asymptotic formulas for trigonometric series, *Indian J. Math.* **9** (1967), 37–41.

[165] *Integrability theorems for trigonometric transforms*, Ergebnisse d. Math., no. 38, Springer-Verlag, 1967.

[166] (editor) *Collected works of Hidehiko Yamabe*, Gordon and Breach, 1967.

[167] 1724 Lilliputians, *Amer. Notes and Queries* **6** (1968), 115–116.

[168] Inequalities for derivatives of polynomials, *Math. Mag.* **42** (1969), 165–174.

[169] Lhospital's rule without mean-value theorems, *Amer. Math. Monthly* **76** (1969), 1051–1053.

[170] *Laplace transforms, characteristic functions, and Lipschitz conditions*, Publ. Ramanujan Inst., no. 1 (1968), 71–74.

[171] Generalized Taylor series, quadrature formulas, and a formula by Kronecker, *SIAM Rev.* **12** (1970), 116–119.

[172] Nejednakosti za nizove polinom, *Matematicka Biblioteka*, sv. **42** (1969), 5–10. Modified version, in Serbo-Croatian, of "Inequalities for derivatives of polynomials," *Math. Mag.* **42** (1969), 165–174.

[173] The Skewes number, *Delta* **1**(4) (1970), 32–36.

[174] Some integral inequalities related to Hardy's inequality, *J. Analyse Math.* **23** (1970), 53–63.

[175] The Jensen-Steffensen Inequality, *Univ. Beograd. Publ. Elektrotehn. Fak. Ser. Mat. Fiz.*, no. 302–319 (1970), 1–8.

[176] Bourbaki, in *Dictionary of Scientific Biography, Volume II*, Scribner's, 1970, pp. 351–353.

[177] (with R. Askey) Some integrability theorems for power series with positive coefficients, *Mathematical Essays Dedicated to A. J. Macintyre*, Ohio University Press, 1970, pp. 23–32.

[178] Calculus as an experimental science, *Amer. Math. Monthly* **78** (1971), 664–667; reprinted in *Two-Year College Math. J.* **2**(1) (1971), 36–39.

[179] (with C. Stutz) Estimating sums with integrals, *Amer. J. Phys.* **39** (1971), 745–753.

[180] (with J. W. Wrench, Jr.) Partial sums of the harmonic series, *Amer. Math. Monthly* **78** (1971), 864–870.

[181] Signs of derivatives and analytic behavior, *Amer. Math. Monthly* **78** (1971), 1085–1093.

[182] A uniqueness theorem for harmonic functions, *J. Approx. Theory* **5** (1972), 425–427.

[183] Summation formulas and band-limited signals, *Tôhoku Math. J.* **24**(2) (1972), 121–125.

[184] Anomalous cancellation, *Two-Year College Math. J.* **3**(2) (1972), 21–24.

[185] The integrability class of the sine transform of a monotonic function, *Studia Math.* **44** (1972), 365–369.

[186] (with H. Pollard) Continuous analogues of series, *Amer. Math. Monthly* **80** (1973), 18–25.

[187] (with A. R. Reddy) Zeros of the successive derivatives of entire functions, *Bull. Amer. Math. Soc.* **79** (1973), 64–65; *J. Math. Anal. Appl.* **42** (1973), 466–473; **49** (1975), 527.

[188] Cantilevered books, *Amer. J. Phys.* **41** (1973), 715.

[189] (with M. Marcus) Inequalities involving a function and its inverse, *SIAM J. Math. Anal.* **4** (1973), 585–591.

[190] (with R. D. Anderson, et al.) Report of the committee on new priorities for undergraduate education in the mathematical sciences, *Amer. Math. Monthly* **81** (1974), 984–988.

[191] Comment on "Kinematics problem for joggers," *Amer. J. Phys.* **42** (1974), 695.

[192] (with M. Marcus) Generalizations of Young's inequality, *J. Math. Anal. Appl.* **46** (1974), 36–40.

[193] (with M. Marcus) Inverse functions and integration by parts, *Amer. Math. Monthly* **81** (1974), 760–761.

[194] (editor) G. Pólya, *Collected Papers*, vols. 1 and 2, MIT Press, 1974.

[195] (with M. L. Boas) A new use for an old counterexample, *Amer. Math. Monthly* **82** (1975), 481–486.

[196] (with C. O. Imoru) Elementary convolution inequalities, *SIAM J. Math. Anal.* **6** (1975), 457–471.

[197] Growth of partial sums of divergent series, *Math. Comp.* **31** (1977), 257–264.

[198] Partial sums of infinite series, and how they grow, *Amer. Math. Monthly* **84** (1977), 237–258.

[199] (with M. S. Klamkin) Extrema of polynomials, *Math. Mag.* **50** (1977), 75–78.

[200] Distribution of digits in integers, *Math. Mag.* **50** (1977), 198–201.

[201] Some unusual definite integrals, *Delta* **7** (1977), 72–76.

[202] Letter to the editor, *Physics Today* **30**(11) (1977), 15.

[203] Estimating remainders, *Math. Mag.* **51** (1978), 83–89.

[204] Extremal problems for polynomials, *Amer. Math. Monthly* **85** (1978), 473–475.

[205] Means derived from convergent series, *Math. Gaz.* **62** (1978), 301.

[206] Zeros of successive derivatives of a function analytic at infinity, *Univ. Beograd. Publ. Elektrotehn. Fak. Ser. Mat. Fiz.*, no. 602–633 (1978), 51–52.

[207] (with H. Pollard and D. V. Widder) The asymptotic behavior of derivatives, *Amer. Math. Monthly* **85** (1978), 749–750.

[208] Travelers' surprises, *Two-Year College Math. J.* **10** (1979), 82–88.

[209] Inequalities for a collection, *Math. Mag.* **52** (1979), 28–31.

[210] Review of Grosswald, Emil, *Bessel polynomials*, in *Bull. Amer. Math. Soc. (N.S.)* **1** (1979), 799–800.

[211] Snowfalls and elephants, pop bottles and π, *Arts and Sciences* [Northwestern University] **2**(1) (1979), 2–5. Reprinted in *Two-Year College Math. J.* **11** (1980), 82–89; *Math. Teacher* **74** (1) (1981), 49–55.

[212] (with C. L. Prather) Final sets for operators on finite Fourier transforms, *Houston J. Math.* **5** (1979), 29–36.

[213] (with G. T. Cargo) Level sets of derivatives, *Pacific J. Math.* **83** (1979), 37–44.

[214] Some remarkable sequences of integers, in *Mathematical Plums*, Math. Assoc. of America, 1979, pp. 38–61.

[215] Anomalous cancellation, in *Mathematical Plums*, Math. Assoc. of America, 1979, pp. 113–129.

[216] Convergence, divergence, and the computer, in *Mathematical Plums*, Math. Assoc. of America, 1979, pp. 151–159.

[217] The Skewes number, in *Mathematical Plums*, Math. Assoc. of America, 1979, pp. 171–179.

[218] Are mathematicians unnecessary? *Math. Intelligencer* **2**(4) (1979), 172–173.

[219] Generalizations of the 64/16 problem, *J. Recreational Math.* **12**(2) (1979), 116–118.

[220] Award for distinguished service to Otto Neugebauer, *Amer. Math. Monthly* **86** (1979), 76–78.

[221] Can we make mathematics intelligible? *Amer. Math. Monthly* **88** (1981), 727–731.

[222] Who needs those mean-value theorems, anyway? *Two-year College Math. J.* **12** (1981), 178–181.

[223] Long multiplication, *California MathematiCs* **7**(1) (1982), 13–14.

[224] Boxing the chain rule, *California MathematiCs* **7**(1) (1982), 36.

[225] Power series for practical purposes, *Two-Year College Math. J.* **13** (1982), 191–195.

[226] Names of functions: The problems of trying for precision, *Math. Mag.* **56** (1983), 175–176.

[227] The versed of Boas, *Two Year College Math. J.* **14** (1983), 342–344.
[228] Mathematics, in *Encyclopædia Britannica Book of the Year*, 1946–49, and for "Ten Eventful Years."
[229] How to publish mathematics, *AWM Newsletter* **14** (1984), 9–12.
[230] Letter to the editor, *Nature* **309** (1985), 10.
[231] (with M. L. Boas) A remark on principal value integrals, *Amer. J. Phys.* **52** (1984), 276.
[232] Inverse functions, *College Math. J.* **16** (1985), 42–47.
[233] (with M. L. Boas) Simplification of some contour integrations, *Amer. Math. Monthly* **92** (1985), 212–213.
[234] Letter to the editor, *Focus* **5**(4) (1985), 4.
[235] (with J. J. Rotman) Letter to the editor, *Amer. Math. Monthly* **92** (1985), 374–375.
[236] Review of Smith, Kennan T., *Primer of modern analysis*, in *SIAM Review* **27** (1985), 93–94.
[237] Letter to the editor, *Math. Intelligencer* **7**(3) (1985), 6, 9.
[238] Computers are icumen in, *Amer. Math. Monthly* **93** (1986), 389.
[239] Counterexamples to L'Hôpital's rule, *Amer. Math. Monthly* **93** (1986), 644–645.
[240] Does "holds water" hold water? *College Math. J.* **17** (1986), 341.
[241] Bourbaki and me, *Math. Intelligencer* **8**(4) (1986), 84–85.
[242] Fishing bayside with horse and wagon, *Capeweek, Cape Cod Times Magazine*, 29 August 1987, p. 6.
[243] (with J. L. Brenner) The asymptotic behavior of inhomogeneous means, *J. Math. Anal. Appl.* **123** (1987), 262–264.
[244] Review of Henrici, Peter, *Applied and computational complex analysis, volume 3*, in *SIAM Review* **29** (1987), 138.
[245] Selected topics from Pólya's work in complex analysis, *Math. Mag.* **60** (1987), 271–274.
[246] *Invitation to Complex Analysis*, Random House, 1987.
[247] Pólya's work in analysis, *Bull. London Math. Soc.* **19** (1987), 576–583.
[248] Memories of bygone meetings, in *A Century of Mathematics in America, Part I*, American Mathematical Society, 1988, pp. 93–95.
[249] Review of Körner, T. W., *Fourier analysis*, in *SIAM Review* **30** (1988), 665–666.
[250] (with A. M. Trembinska) An extension of Carlson's theorem for analytic functions, *J. Math. Anal. Appl.* **129** (1988), 131–133.
[251] Letter to the editor, *Pi Mu Epsilon J.* **8** (1988), 521–522.
[252] The sign of $e^{\pi} - \pi^{e}$, *Pi Mu Epsilon J.* **8** (1988), 521.
[253] (with H. P. Boas) Short proofs of three theorems on harmonic functions, *Proc. Amer. Math. Soc.* **102** (1988), 906–908.

[254] Indeterminate forms revisited (videocassette), American Mathematical Society, 1989 (AMS-MAA joint invited address, Phoenix, January 1989).

[255] When is a C^∞ function analytic? *Math. Intelligencer* **11**(4) (1989), 34–37.

[256] Multiplying long numbers, *Math. Mag.* **62** (1989), 173–174.

[257] Ralph P. Boas Jr., in *More Mathematical People* (edited by D. J. Albers et al.), Harcourt Brace Jovanovich, 1990, pp. 23–41.

[258] Indeterminate forms revisited, *Math. Mag.* **63** (1990), 155–159.

[259] George Pólya/December 13, 1887–September 7, 1985, *Biographical Memoirs, National Academy of Sciences* **59** (1990), 338–355.

[260] (editor) *A. J. Lohwater's Russian-English Dictionary of the Mathematical Sciences, 2nd ed.*, American Mathematical Society, 1990.

[261] Otto Neugebauer, 1899–1990, *Notices Amer. Math. Soc.* **37**(5) (1990), 541.

[262] (editor) S. H. Gould, *A Manual for Translators of Mathematical Russian*, American Mathematical Society, 1991.

[263] What St. Augustine didn't say about mathematicians, *Pi Mu Epsilon J.* **9** (1991), 309. An earlier version appeared in the *Amer. Math. Monthly* **86** (1979), 138.

[264] Naming things, *Math. Mag.* **66** (1993), 43.

Translations

[1] Levin, B. Ya., *Distribution of zeros of entire functions*, revised edition, American Mathematical Society, 1980 (translated with J. M. Danskin, F. M. Goodspeed, J. Korevaar, A. L. Shields, and H. P. Thielman).

[2] Markushevich, A. I., *Introduction to the classical theory of Abelian functions*, American Mathematical Society, 1992 (translation by G. Bluher, translation edited by Boas).

[3] Nikiforov, A. F., and V. B. Uvarov, *Special functions of mathematical physics*, Birkhäuser, 1988.

[4] Shiryayev, A. N., *Probability*, 2nd edition, Springer-Verlag, to appear.

[5] Suetin, P. K., *Polynomials orthogonal over a region and Bieberbach polynomials*, American Mathematical Society, 1974.

[6] Vorob'ev, N. N., *Foundations of game theory*, Birkhäuser, 1994.

[7] Voronovskaya, E. V., *The functional method and its applications*, American Mathematical Society, 1970.

INDEX